THE CHEMICAL SOCIETY
MONOGRAPHS FOR TEACHERS No. 26

Elements of
Organometallic Chemistry

F. R. HARTLEY, MA, DPhil, ARIC

Department of Chemistry, The University of Southampton

540
H332e

LONDON: THE CHEMICAL SOCIETY

Monographs for Teachers

This is another publication in the series of Monographs for Teachers which was launched in 1959 by the Royal Institute of Chemistry. The initial aim of the series was to present concise and authoritative accounts of selected well-defined topics in chemistry for those who teach the subject at GCE Advanced level and above. This scope has now been widened to cover accounts of newer areas of chemistry or of interdisciplinary fields that make use of chemistry. Though intended primarily for teachers of chemistry, the monographs will doubtless be of value also to a wider readership, including students in further and higher education.

© *The Chemical Society, 1974*

First published October 1974

Published by The Chemical Society, Burlington House, London W1V 0BN, and distributed by The Chemical Society Publications Sales Office, Blackhorse Road, Letchworth, Herts SG6 1HN

Printed by Adlard & Son Ltd, Bartholomew Press, Dorking

Preface

The idea of this monograph arose during a one-day symposium for schoolteachers on 'Organometallic Chemistry' that was held in Southampton in October 1971. A number of schoolteachers approached me and pointed out that whilst what the speakers had been saying was very interesting there were two major stumbling blocks to introducing organometallic chemistry into school teaching. The first was the lack of a suitably simple textbook and the second was the lack of any suitable experiments. Over the past three years, with a great amount of help from schoolteachers in the Southampton area and colleagues in the university, especially Dr C. Burgess, I have attempted to remedy this situation by offering this monograph, in which I have included a number of experiments that may be used to illustrate some of the principles of practical organometallic chemistry.

Many organometallic experiments use sophisticated equipment to overcome the inherent instability of organometallic compounds, and are thus difficult to carry out in schools. The experiments given here have all been extensively modified from the original literature and in this I am indebted to Mrs G. Temple-Nidd of the Southampton College for Girls, Mr T. G. B. Hackston of Barton Peveril College (Eastleigh), Dr A. W. Shaw of the South Wilts Grammar School for Girls (Salisbury) and Dr R. P. Smith of Richard Taunton College (Southampton) all of whom have devoted a considerable amount of their free time to developing these experiments. I am also indebted to the Goldsmiths' Company who generously gave us some financial help for this work from their John Perryn charity bequest.

Organometallic chemistry is a relatively recent subject. Although the first organometallic compound was reported in 1830 and the foundations of non-transition metal organometallic chemistry were laid towards the end of the nineteenth and at the beginning of the twentieth century by Grignard and others, transition metal organo-metallic chemistry did not really get under way until the 1950s when it was spurred on by the discovery of ferrocene. Recently it has received further recognition through the award of the 1973 Nobel Prize for Chemistry to two of the leading organometallic chemists, namely Professors G. Wilkinson and E. O. Fischer. Whilst Grignard reagents have long been a part of university undergraduate courses, transition metal organometallic chemistry has only been taught widely in the past eight years. Hence there are many schoolteachers who have never been formally taught anything about the subject and some of these may well ask, 'why teach organometallic chemistry in schools?' My answer is, firstly organometallic chemistry has

iii

recently become a very important part of the armoury of the industrial chemist and is thus not a further branch of chemistry of purely academic interest, and secondly it lies on the boundaries of organic and inorganic chemistry and thus utilizes the concepts of both, helping to fuse the subject together and break down the traditional barriers which it was so fashionable to build a few years ago. Throughout this monograph and especially in chapter 5 I have, where possible, emphasized the applications of organometallic chemistry.

Whilst it would be impertinent of a university chemist to tell schoolteachers how to teach their subject, some suggestions as to how organometallic chemistry could be introduced into a curriculum may be in order. In the first place students taking the Nuffield scheme undertake a two-stage preparation around the end of their first year. Some of the experiments described in Chapter 7 could well be used here. Looking further to the future, organometallic chemistry could well join chemical engineering, biochemistry and food science as one of the areas of the subject for further specialization.

Contents

CONTENTS

1. Introduction

What is an organometallic compound?

For the purposes of this monograph an organometallic compound is considered to be any compound containing a direct bond between a metal atom and one or more carbon atoms. However, metal–carbonyls (M—CO), metal cyanides (M—CN) and metal carbides (*e.g.* CaC_2) are not considered in this monograph because their properties are closely related to those of many other compounds usually considered in inorganic chemistry.

Types of organometallic compounds

The first useful subdivision of organometallic compounds is by metal which gives five types of organometallic compound: ionic, covalent, electron deficient, transition metal and others.

Ionic

Ionic organometallic compounds are generally only formed by the most electropositive elements, namely sodium, potassium, rubidium, caesium and francium. However, slightly less electropositive metals such as magnesium can occasionally form ionic compounds when the negative charge on the organic moiety can be delocalized over several carbon atoms. Thus sodium forms ionic butylsodium whereas the less electropositive magnesium forms a covalent butyl compound. Magnesium, however, forms an ionic cyclopentadienide ($Mg^{2+}\{C_5H_5\}_2$) where the negative charge can be delocalized over the five carbon atoms of the cyclopentadienide ring.

Ionic organometallic compounds are typical ionic compounds and are consequently insoluble in hydrocarbon solvents and crystalline solids when obtained pure. This, however, is not easy as they are readily hydrolysed and sensitive to oxygen.

Covalent

Volatile, covalent organometallic compounds are formed by zinc, cadmium and mercury and the non-transition metals of groups III (other than aluminium), IV and V. They are the simplest type of organometallic compound known involving an electron-pair bond to which the metal and the organic group each contribute a single electron. Their properties resemble those of typical organic compounds in that they are volatile, soluble in organic solvents and insoluble in water. They include, of course, compounds such as tetraethyllead the famous (or infamous!) antiknock additive in petrol.

1

Electron deficient

The organometallic compounds of lithium, beryllium, magnesium and aluminium all form electron deficient structures, in which there are insufficient valence electrons to allow all the atoms to be linked by traditional two-electron two-centre bonds.   All four metals form very strongly polarizing cations, that is they have a very high 'charge to radius' ratio, which enables the cations to 'pull electron density away from an anion' (*i.e.* polarize that anion).   As a result the charge separation in the potentially ionic methyl compound

$$M^+ \quad {}^-CH_3$$

decreases and a covalent, but polar, structure results.

$$\overset{\delta+}{M}\text{---}\overset{\delta-}{CH_3}$$

It is apparent that such a highly polar structure will associate strongly so that we may expect the overall structure to be polymeric.   This is just what does happen.   Additionally, there is an interesting gradation in properties in that aluminium alkyls form dimers, beryllium and magnesium alkyls form linear chains and lithiuma lkyls form 3-dimensional polymers   These different structures are reflected in the volatilities of these compounds which increase sharply from methyllithium through dimethylberyllium to trimethylaluminium (Table 1) in spite of an increase in the molecular weight of the monomer.

Table 1.   Properties of electron deficient compounds.

Compound	Molecular weight of monomer	Structure	Volatility
LiMe	21.96	3d-polymer	Infusible
BeMe$_2$	39.08	Linear chain	Sublimes at 200°C
AlMe$_3$	72.09	Dimer	Melts at 15.4°C

Transition metal

The most important difference between transition metals and non-transition metals is the ability of the former to bond to more than one carbon atom.   This ability, which stems from the symmetry properties of the *d*-orbitals present in the valence shells of transition metals, enables transition metals to form such a wide range of complexes that it is more convenient to classify them in terms of the organic ligand (p 3) than in terms of the metal.

The lanthanide and actinide metals, although their organometallic chemistry has not been studied in such detail as the transition metals, appear to resemble the latter in their ability to bond to more than one carbon atom of an organic ligand, *e.g.* [U(C$_8$H$_8$)$_2$] (I)

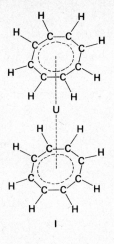

I

Others

The organometallic compounds of calcium, strontium, barium and radium have not been widely investigated, possibly because the metals are relatively inaccessible and their alkyls show no significant advantages over those of magnesium and lithium reagents. In general their properties appear to be intermediate between those of magnesium alkyls and the alkali metal alkyls.

Classification of ligands

Whilst non-transition metals form organometallic compounds with only a narrow range of ligands such as alkyl, aryl and σ-cyclopentadienyl groups, transition metals form complexes with a very wide range of organic groups. Accordingly it is valuable to classify these compounds according to ligand type rather than metal. The most important difference between transition and non-transition metals is the ability of the former to bond to more than one carbon atom of an organic ligand and it is convenient to classify ligands according to the number of their carbon atoms which are involved in bonding to the metal.*

---

* The classification obtained here from the number of carbon atoms bound to the metal is essentially the same as that developed by Green (in *Organometallic compounds*, Vol. 2 — *see* Suggested Reading), on the basis of the number of electrons donated by the ligand. The present classification is offered not simply to be different but in the hope that it may be conceptually simpler to grasp. Furthermore, it enables us to retain the concept of the formal oxidation state (pp 7–8) that has been of great value in inorganic chemistry.

One-carbon-bonded ligands

One-carbon-bonded ligands are ligands in which only one carbon atom of the ligand is bound directly to the metal. Such ligands can be further subdivided into three groups, namely hydrocarbon, acyl and carbene ligands.

Hydrocarbon ligands include alkyl ($-CH_3$), aryl ($-C_6H_5$), alkenyl ($-CR=CR_2$), σ-cyclopentadienyl ($-C_5H_5$), and alkynyl ($-C\equiv CR$) groups. All these groups form compounds with both transition and non-transition metals.

Acyl ligands involve the direct bonding of the carbon atom of an acyl group to a metal

Such complexes are formed by transition but not by non-transition metals.

Carbene ligands involve the coordination of the carbon atom of a carbene

to a metal. The metal is generally a transition metal and the groups X and Y are often groups such as $-OR$ and $-NR_2$.

Two-carbon-bonded ligands

Alkenes and alkynes act as two-carbon-bonded ligands because the carbon atoms at each end of the multiple bond are both involved in bonding to the metal atom.

It should be noted that the two carbon atoms need not be equidistant from the metal atom, and indeed in many, if not all, complexes of unsymmetrically substituted alkenes and alkynes such as $RCH=CH_2$, the two metal–carbon bond lengths are slightly different. As already noted alkenes and alkynes only form complexes with transition metals.*

Three-carbon-bonded ligands

Complexes in which 3-carbon atoms are bound directly to a transition metal are known as π-allyl or π-enyl complexes. The simplest of these ligands is the π-allyl (π-C_3H_5) group but all the carbon atoms may bear substituents, or the three carbon atoms may be part of a ring. The complexes are normally represented schematically by

and should be distinguished from the σ-allyl complexes in which the allyl group is bound as a one-carbon-bonded ligand ($M-CH_2-CH=CH_2$). Further examples of π-allylic ligands are:

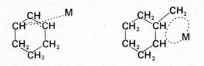

Four-carbon-bonded ligands

There are two major types of four-carbon-bonded ligands, (a) acyclic ligands such as butadiene and (b) cyclic ligands such as cyclobutadiene.

Whilst cyclobutadiene always acts as a four-carbon-bonded ligand butadiene does not always do so. Occasionally it prefers to act as a two-carbon-bonded ligand to one (II) or two (III) metal atoms.

---

* These complexes are known as π-complexes a label that results from their bonding (p 40). They should be distinguished from the σ-complexes ($M-CR=CR_2$ and $M-C\equiv CR$) formed by alkenes and alkynes that have lost one hydrogen atom from the carbon atom adjacent to the metal. These latter are, of course, classified above under hydrocarbon ligands.

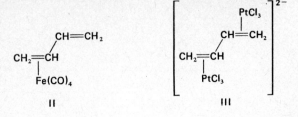

II   III

Five-carbon-bonded ligands

Ligands in which five or more carbon atoms are directly bound to a transition metal are generally cyclic ligands. Five-carbon-bonded ligands are known as '-dienyl' ligands and although typified by the 5-membered ring ligand 'π-cyclopentadienyl' are also formed by 6- and 7-membered hydrocarbon ligands. The π-cyclopentadienyl group gives rise to the famous metallocene series of complexes — for example, ferrocene, nickelocene and ruthenocene — in which a divalent transition metal is sandwiched between two planar cyclopentadienyl rings (IV). $[(\pi\text{-}C_6H_7)Fe(CO)_3]BF_4$ (V) is an example of a five-carbon-bonded complex formed by the 6-membered cyclohexadienyl ring.

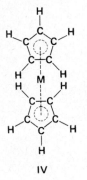

IV

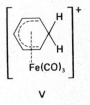

V

Six-carbon-bonded ligands

Although benzene is undoubtedly the most important six-carbon-bonded ligand other ligands such as cycloheptatriene and cyclo-octatetraene also form complexes in which only six of their carbon atoms are bound directly to the metal. Heterocyclic ligands such as pyridine and thiophene also form complexes which are electronically and structurally analogous to benzene complexes.

Seven-carbon-bonded ligands

π-Cycloheptatrienyl complexes, (VI), in which seven carbon atoms are bound to the transition metal are formed by vanadium, chromium, manganese, iron, cobalt, molybdenum, tungsten and rhodium.

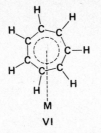

VI

Eight-carbon-bonded ligands

Cyclooctatetraene (VII) forms a number of complexes particularly

VII

with the actinides in which it is planar with all eight carbon atoms equidistant from the metal (*e.g.* $U(C_8H_8)_2$ — *see* p 3). However, cyclooctatetraene also forms complexes in which it acts as a 4- (VIII), 2×4- (IX), 4×2- (X) and a 6-carbon-bonded ligand (XI).

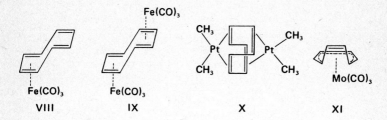

| VIII | IX | X | XI |

Oxidation state

Although the oxidation state of an element is only a formalism that is obtained by applying a set of rules and thus has no absolute meaning it has nevertheless become a very useful formalism. The oxidation state of a metal is defined as 'the charge remaining on that metal after all the ligands have been removed in their closed shell conditions (*e.g.* chlorine as Cl^-) and any metal–metal bonds have been cleaved homolytically — *i.e.* the presence of a metal–metal bond does not alter the oxidation state of the metal'. Thus magnesium in $MgCl_2$ is Mg^{2+} after removing chlorine as Cl^- and so magnesium is said to be in the +2 oxidation state which is written as magnesium(II). Similarly in $[(PPh_3)_2PtCl_2]$ after removing Cl as Cl^- and triphenylphosphine as the neutral PPh$_3$ ligand $+2$ is left as the oxidation state of platinum — *i.e.* platinum(II).

How do these rules apply to organometallic compounds? The even-carbon-bonded ligands such as ethene, butadiene and benzene are all neutral in their closed shell configurations and so do not influence the oxidation state of the metal. By contrast the odd-carbon-bonded ligands are all formally negatively charged in their closed shell configuration and so contribute one unit to the oxidation state of the metal. Thus to get eight electrons in the sp^3 orbitals around the carbon of methyl we require four electrons from the carbon valence orbitals, three from the hydrogen atoms and one more from the negative charge. Thus methyl is formally CH$_3^-$

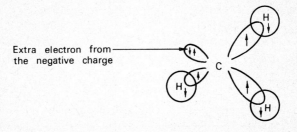

Extra electron from the negative charge

Applying these rules to a number of organometallic compounds.

1. MeMn(CO)$_5$ Carbon monoxide is neutral.
        Methyl carries formal 1− charge.
        ∴   Manganese is in +1 oxidation state.

2. K$^+$[Pt(C$_2$H$_4$)Cl$_3$]$^-$ Ethene is neutral.
        Chlorine carries formal 1− charge.
        Complex ion carries overall 1− charge.
        ∴   Platinum is in +2 oxidation state.

3.

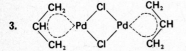

π-Allyl is an odd-carbon-bonded ligand and therefore carries a formal 1− charge. Chlorine carries a formal 1− charge. Thus two palladium atoms share a formal 4+ charge and so palladium is in +2 oxidation state.

4.

Cyclobutadiene is an even-carbon-bonded ligand and is therefore formally neutral. Carbon monoxide is neutral.
∴   Iron is in zero oxidation state.

As a further help in assigning the formal charge on ligands it may be noted that ligands whose name ends in -ene are even-carbon-bonded ligands and formally neutral (*e.g.* ethene, benzene *etc*), whereas ligands whose name ends in -yl are odd-carbon-bonded ligands and formally uninegative (*e.g.* methyl, π-allyl, π-cyclo-heptatrienyl *etc*).

Stability

In order to describe a compound as 'stable' it is important to specify what the compound is stable to, for example, is it stable to heat, air oxidation or hydrolysis? Consider a Grignard reagent such as methylmagnesium iodide. It is clearly stable at around room temperature with respect to decomposition to magnesium and methyl iodide, after all it is formed by the interaction of these two, but it is very sensitive to oxidation and hydrolysis (reactions 1 and 2),

$$2CH_3MgI + O_2 \rightarrow 2CH_3OMgI \qquad\qquad 1$$

$$CH_3MgI + H_2O \rightarrow CH_4(g) + Mg(OH)I \qquad\qquad 2$$

so that it is necessary to exclude air and moisture during its preparation and subsequent storage. Clearly then to describe such a compound as 'stable' without further qualification is meaningless.

There are basically two types of stability, namely thermodynamic and kinetic.

Thermodynamic stability

A compound is thermodynamically stable if 'its free energy is lower (*i.e.* more negative) than the sum of the free energies of the products formed when it reacts' (Scheme 1). Unfortunately there is not a lot of

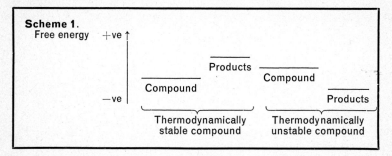

Scheme 1.

free energy data available for organometallic compounds although there is rather more enthalpy data — generally obtained from measuring heats of combustion. If this is to be used then allowance must be made for the entropy term in the equation

$$\Delta G = \Delta H - T\Delta S$$

Table 2. Estimates of the thermodynamic stabilities of selected organometallic compounds.

Compound	Type of stability to be considered	Enthalpy data (ΔH°_f, kJ mol^{-1})	Net enthalpy change	Estimate of entropy	Conclusion
Me$_2$Cd	Thermal stability, decomposition assumed to occur by Me$_2$Cd → Cd + C$_2$H$_6$	Me$_2$Cd(l) +70.0 C$_2$H$_6$(g) −84.5 Cd(cryst) 0	−154.5 kJ mol^{-1}, i.e. favours decomposition	Will favour decomposition since a gas is formed	Thermodynamically unstable to heat
EtLi	Thermal stability, decomposition assumed to occur by EtLi → LiH + CH$_2$=CH$_2$	EtLi(cryst) −58.55 C$_2$H$_4$(g) +52.40 LiH(cryst) −90.45	+20.50 kJ mol^{-1}, i.e. favours stability	Will favour decomposition because the entropy of gaseous ethene is high (+65.4 kJ mol^{-1} at room temperature) compared to the entropies of crystalline solids (e.g. entropy of formation of LiH = +7.55 kJ mol^{-1} at room temperature)	Thermodynamically unstable to heat
Cr(C$_6$H$_6$)$_2$	Thermal stability, decomposition assumed to give Cr metal and benzene	Cr(C$_6$H$_6$)$_2$(cryst) +150.80 Cr(cryst) C$_6$H$_6$(l) +49.05	−52.7 kJ mol^{-1}, i.e. favours decomposition	Will favour decomposition since a liquid (or gas, above its boiling point) is formed	Thermodynamically unstable to heat

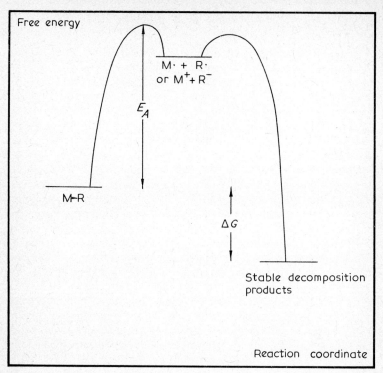

FIG. 1. The decomposition of a metal–alkyl (M−R) either homolytically (giving M · + R ·) or heterolytically (giving M$^+$ + R$^-$).   The larger E_A the more kinetically stable M−R will be.

(ΔG = free energy change, ΔH = enthalpy change, ΔS = entropy change and T = absolute temperature).   Except in a few special cases this allowance need not be sophisticated, but can rather take account of the fact that the entropies of gases are much larger than the entropies of liquids which in turn are much larger than the entropies of solids.   The stabilities of three compounds are analysed in this way in Table 2.

Kinetic stability

Why is it that, although the compounds in Table 2 are thermodynamically unstable, they can be isolated?   The reason is, of course, that for a reaction to proceed **two** criteria must be fulfilled:

(*i*) There must be a favourable free energy change.
(*ii*) There must be a pathway of sufficiently low activation energy for it to occur at a measurable rate.   If no such pathway exists, the reaction will not proceed and even a thermodynamically unstable

2

compound may be kinetically stable.   This is shown schematically in *Fig. 1*.

It should be emphasized that the intermediate products formed from organometallic compounds in which the metal is bound to an odd number of carbon atoms (1, 3, 5, *etc* pp 3–7) are extremely reactive organic intermediates such as free radicals or carbanions. These reactive fragments readily form stable products, for example by dimerization, giving the thermodynamic driving force for the reaction.   These end-products do not normally react with the free metal to reform the metal–alkyl and so the reaction is irreversible.   Thus the preparation of stable organometallic compounds of 1-, 3-, 5- and 7-carbon-bonded fragments depends on designing a complex in which the activation energy (E_A of *Fig. 1*) is large.   The recent explosion in organometallic chemistry is a reflection of the fact that organometallic chemists have now discovered ways of doing this.

In contrast to organometallic compounds where the metal is directly bound to an odd number of carbon atoms, the organic fragments formed by the decomposition of organometallic compounds in which even numbers of carbon atoms are bound to the metal are generally stable entities such as ethene, butadiene and benzene and so decomposition can now be reversible.   These compounds thus resemble typical inorganic complexes in which ligands such as chloride ions or amines can be dissociated reversibly. Thus in the same way that in aqueous solution equilibrium 3

$$[PtCl_4]^{2-} + H_2O \rightleftharpoons [PtCl_3(H_2O)]^- + Cl^- \qquad 3$$

lies to the right in the absence of added chloride ion and to the left in the presence of an excess of chloride ion, so the complex [PtCl₃-(alkene)]⁻ can in the presence of chloride ions reversibly lose alkene (reaction 4).

$$[PtCl_3(alkene)]^- + Cl^- \rightleftharpoons [PtCl_4]^{2-} + alkene \qquad 4$$

Two of the main properties of a complex that lead to a low activation energy for reaction are the presence of low energy empty orbitals and an ability on the part of the metal to expand its coordination number.   Consider carbon tetrachloride.   The valence s and p orbitals on the carbon atom are all involved in bonding to the chlorine atoms (C is sp^3 hybridized) and there are no further empty orbitals of sufficiently low energy to be accessible for bonding. Carbon tetrachloride is consequently stable to water.   The silicon atom in SiCl₄, however, has some unfilled orbitals of relatively low energy in its $3d$ orbitals. Consequently it is possible for water molecules to form intermediate species in which the water molecule donates 2-electrons to the silicon.   It is now possible for the hydrogen on the water and the chlorine on the silicon to react together (reaction 5).

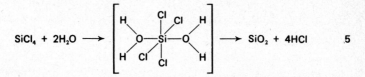

$$SiCl_4 + 2H_2O \longrightarrow \left[\begin{array}{c} \text{structure} \end{array} \right] \longrightarrow SiO_2 + 4HCl \qquad 5$$

Thus by having low energy unfilled $3d$ orbitals and so being able to coordinate the water molecules the silicon has been able to split the large activation energy for the total reaction into two smaller activation energies, one for coordination of water and the other for elimination of hydrogen chloride from the intermediate $[SiCl_4(H_2O)_2]$, both of which are readily attainable at room temperature.

The methyl groups in tetramethylsilane ($SiMe_4$) are less electronegative than chlorine and so do not attract electron density away from the silicon to the same extent as chlorine. As a result the $3d$ orbitals are of higher energy in $SiMe_4$ than in $SiCl_4$ and so are less accessible in $SiMe_4$. Consequently the silicon in $SiMe_4$ does not readily coordinate extra ligands and so it does not react with water. Tetramethylsilane is kinetically stable to thermal decomposition, oxidation and hydrolysis, properties that are exploited in its use as the standard internal reference for chemical shifts in proton nmr work.

Compounds of transition metals often have empty valence shell orbitals available and may also have vacant sites — in which case they are said to be 'coordinatively unsaturated'. It is thus no accident that most of the stable compounds involve either bulky ligands which prevent reagents approaching the metal closely, or else have electron configurations in which the bonding and non-bonding molecular orbitals are filled and the antibonding molecular orbitals are empty.*

In general an examination of the literature reveals that there are very few thermodynamically stable organometallic compounds. Organometallic chemistry is thus dependent on the phenomenon of kinetic stability.

---

* This provides the basis of the 16- and 18-electron rules for stable organometallic compounds discussed by C. A. Tolman, *Chem. Soc. Rev.*, 1972, **1**, 337.

2. Preparation of Organometallic Compounds

Practical considerations

Before looking at specific routes for the preparation of organometallic compounds some of the general practical considerations will be considered. Many organometallic compounds are unstable to oxidation and hydrolysis, and must, therefore, be prepared in the absence of oxygen in dry apparatus using solvents and reagents that have been rigorously dried. In many cases where the final organometallic compound is stable to oxygen and water, the intermediates used in its preparation are not — for example Grignard reagents and alkyllithium compounds are both sensitive to oxygen and water — and so again oxygen and water must be excluded. Oxygen is generally excluded using 'white-spot' nitrogen (the trade name for oxygen-free nitrogen) and water, in the form of moisture from the air, can be excluded either by using guard-tubes filled with silica gel or phosphorus pentoxide or, if the reaction is to be carried out under nitrogen, by ensuring that there is always a positive pressure of nitrogen present. This is conveniently effected using a guard tube such as that shown in *Fig. 2* to vent the nitrogen to the air.

A second practical consideration which must be borne in mind is that, whilst the yields of some organometallic compounds are very high — some as high as 100 per cent — the yields of many organometallic compounds are low. They can sometimes be so low that, although a reaction is written in a particular way in the literature, the equation does not describe the way in which the bulk of the reaction occurs. In such cases what the equation describes are the readily isolated products of the reaction, for example the formation of a precipitate, a gas or a product that is readily extracted from the reaction mixture into an immiscible solvent.

Why are organometallic compounds formed?

At the end of Chapter 1 it was concluded that most, if not all, organometallic compounds are thermodynamically unstable with respect to thermal decomposition. In addition, a considerable proportion of organometallic compounds have positive standard enthalpies of formation. As a result one of the points I shall try to emphasize during the course of this chapter is why, in a given reaction, the organometallic compound is formed at all.

The preparation of organometallic compounds

The preparative routes to organometallic compounds given here have

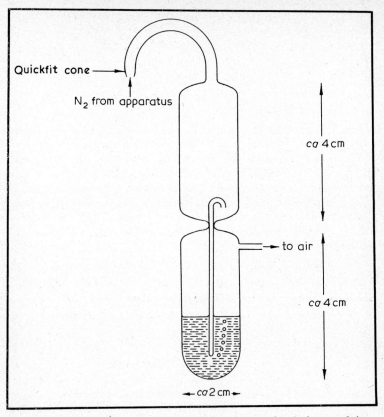

Fig. 2. Guard-tube for venting nitrogen to air and preventing the ingress of air or moisture.

been classified in accordance with the ligand classification given in Chapter 1 (pp 3–7).

One-carbon-bonded hydrocarbon complexes

Organometallic compounds containing one-carbon-bonded hydrocarbon ligands can be prepared by a wide range of reactions, some of the more important of which are described below.

Metal + alkyl halide.   The success of the direct reaction between a metal and an alkyl halide, used for example in the preparation of Grignard reagents (equation 6),

$$C_2H_5Br + Mg \rightarrow C_2H_5MgBr \qquad\qquad 6$$

depends not on the enthalpy of formation of the metal–alkyl bond but rather on that of the metal–halide. Thus in the formation of ethyl-lithium (equation 7)

$$C_2H_5Cl(g) + 2Li(s) \rightarrow LiEt(s) + LiCl(s)$$

$$\Delta H_f^{\ominus} \text{ (kJ mol}^{-1}) \quad -105 \qquad 0 \qquad -58 \qquad -409$$

$$\Delta H_{\text{reaction}}^{\ominus} = -362 \text{ kJ mol}^{-1} \qquad 7$$

the heat of formation of ethyllithium is insufficient to return the energy needed to cleave the carbon–chlorine bond in ethyl chloride. However, the large favourable enthalpy of formation of lithium chloride drives the reaction to the right with a net enthalpy change of -362 kJ mol^{-1}. Since the formation of an exothermic metal halogen bond is essential for this reaction only electropositive metals will normally react directly with alkyl halides. However, on occasion, it is possible to alloy a less electropositive metal with a more electropositive one, the former forming an alkyl and the latter a halide. This is done commercially in the preparation of tetraethyl lead (equation 8).

$$4C_2H_5Cl + 4NaPb(\text{alloy}) \rightarrow Pb(C_2H_5)_4 + 3Pb + 4NaCl \qquad 8$$

A somewhat similar approach is involved when the Wurtz reaction is used to synthesize tetraphenylgermane (equation 8a),

$$4PhBr + GeCl_4 + 8Na \rightarrow Ph_4Ge + 4NaCl + 4NaBr \qquad 8a$$

although in this case germanium remains in the $+4$ oxidation state throughout.

Although the heats of formation of metal halides increase as the Periodic Table is ascended (*i.e.* I$^-$ < Br$^-$ < Cl$^-$ < F$^-$) the reactivities of alkyl halides with metals decrease as the group is ascended. This is a kinetic rather than a thermodynamic effect and arises from the activation energy needed to cleave the carbon–halogen bond. The carbon–halogen bond energies are approximately 440 (C—F), 330 (C—Cl), 276 (C—Br) and 238 kJ mol^{-1} (C—I) and consequently the rates of reaction of alkyl halides increase RI > RBr > RCl $\gg$ RF.

Organometallic + metal halide. The preparation of an organometallic compound by reaction of another organometallic compound with a metal halide depends on the initial organometallic compound being that of a more electropositive metal and the metal halide being that of a less electropositive metal. Typical reactions are:

$$2LiMe + [(PEt_3)_2PtBr_2] \rightarrow [(PEt_3)_2PtMe_2] + 2LiBr \qquad 9$$

$$MeMgBr + [(PEt_3)_2PtBr_2] \rightarrow [(PEt_3)_2PtMeBr] + MgBr_2 \qquad 10$$

$$2LiEt + HgCl_2 \rightarrow HgEt_2 + 2LiCl \quad \Delta H_{\text{reaction}}^{\ominus} = -444 \text{ kJ mol}^{-1} \qquad 11$$

$$-58.5 \quad -230.0 \quad +27.0 \quad -409.0 \quad \Delta H_f^{\ominus} \text{ (kJ mol}^{-1})$$

$$4R_3Al + 4NaCl + 3SnCl_4 \rightarrow 3R_4Sn + 4Na[AlCl_4] \qquad 12$$

and in all cases the more electropositive metal loses its alkyl groups and gains halide groups, thus giving the driving force for the reaction. An enthalpy analysis of reaction 11 shows that both metals would prefer to form chlorides rather than ethyls, but the more electropositive lithium obtains the chloride with the result that mercury ends up with the ethyl groups in what is in fact an endothermic compound. In general alkyl lithium reagents and Grignard reagents react similarly, but where differences are observed, as in reactions 9 and 10, the alkyl lithium reagents are the more reactive, probably due to the slightly greater energy of the lithium–halide bond.   Reaction 12 is interesting in that it is often difficult to get alkylations with trialkyl-aluminium compounds to go to completion.   For example, tin tetrachloride and triethylaluminium yield Et_nAlCl_{3-n} and $SnEt_{3-n}Cl_{n+1}$ ($n = $ 0–3), unless sodium chloride is added, when the reaction goes to completion by virtue of the high free energy of formation of $Na[AlCl_4]$.

This method is probably the most important route to compounds of one-carbon-bonded hydrocarbon ligands, and is certainly the most important route to transition metal complexes of these ligands.

Metal exchange.   When the alkyl of a less electropositive metal is treated with a more electropositive metal the alkyl group is exchanged.   This reaction (reaction 13)

$$HgEt_2 + 2Li \rightarrow Hg + 2LiEt \quad \Delta H^{\ominus}_{reaction} = -144 \text{ kJ mol}^{-1} \qquad 13$$

$$+27.0 \qquad 0 \qquad 0 \qquad -58.5 \quad \Delta H^{\ominus}_f \text{ (kJ mol}^{-1}\text{)}$$

is the first example where the driving force of the reaction is provided by the formation of the organometallic itself.   Mercury alkyls are the commonest alkylating agents used to effect this reaction, because mercury lies low in the electrochemical series, yet its alkyls are readily obtained and are air stable, making handling easy.

Metal carbonylate anion + alkyl halide.   A reaction that is closely allied to the first method is the reduction of a metal carbonyl to a metal carbonylate anion followed by reaction of this with the alkyl halide (reaction 14).

$$Mn_2(CO)_{10} \xrightarrow[\text{Tetrahydrofuran}]{\text{Na/Hg}} Na^+[Mn(CO)_5]^- \xrightarrow{\text{MeI}} [MeMn(CO)_5] + NaI \qquad 14$$

The driving force for this reaction is provided by the formation of insoluble sodium iodide.

Alkene (or alkyne) insertion into an M–X bond.   The general reaction can be represented as 15 or 16,

$$M-X + \quad C=C \quad \rightarrow M-\overset{|}{\underset{|}{C}}-\overset{|}{\underset{|}{C}}-X \qquad 15$$

$$M-X + -C\equiv C- \rightarrow M-\overset{|}{C}=\overset{|}{C}-X \qquad 16$$

where X is usually either H or an alkyl group and M can be a whole range of metals particularly boron, aluminium, silicon and transition metals. With boron the reaction, known as hydroboration, is of vital importance in synthetic organic chemistry because of the wide range of reactions of the boron–alkyl compounds formed, which enable alkenes to be converted to other compounds (*see* H. C. Brown, *Hydroboration*, published by W. A. Benjamin, New York, 1962).

The insertion of alkenes into aluminium–hydrogen bonds (reactions 17–19)

$$Al + 3/2H_2 + 2Et_3Al \rightarrow 3Et_2AlH \qquad 17$$

$$3Et_2AlH + 3C_2H_4 \rightarrow 3Et_3Al \qquad 18$$

giving overall

$$Al + 3/2H_2 + 3C_2H_4 \rightarrow Et_3Al \qquad 19$$

is the principal commercial route to aluminium alkyls — some of the most important commercial alkyls, due partly to their use in the polymerization of alkenes (pp 59–61). The insertion of ethene into an aluminium–alkyl bond is used in the preparation of C_{14} alcohols which are used to prepare biodegradable detergents. Thus insertion (reaction 20) is in competition with elimination (reaction 21), which is of course the reverse of insertion, and at temperatures of about 160 °C an average chain length of about C_{14} is obtained. Oxidation and hydrolysis (reaction 22) yield the desired alcohol.

$$AlEt_3 + C_2H_4 \xrightarrow{100-120°\,C} Al\overset{CH_2CH_2Et}{\underset{Et}{\diagdown Et}} \rightsquigarrow Al\overset{(C_2H_4)_nEt}{\underset{(C_2H_4)_1Et}{\diagdown (C_2H_4)_mEt}} \qquad 20$$

$$Al\overset{CH_2CH_2R}{\underset{Et}{\diagdown Et}} \xrightarrow{>120°\,C} Al\overset{H}{\underset{Et}{\diagdown Et}} + RCH=CH_2 \qquad 21$$

$$Al(C_{14}H_{29})_3 + 3/2O_2 \rightarrow Al(OC_{14}H_{29})_3 \xrightarrow{3H_2O} Al(OH)_3 + 3C_{14}H_{29}OH \qquad 22$$

The insertion of an alkene into a silicon–hydrogen bond, (reaction 23),

$$X_3SiH + H_2C=CHR \rightarrow X_3SiCH_2CH_2R \qquad 23$$

known as hydrosilation, provides the basic raw materials necessary for silicone manufacture.

Perhaps the most important feature of the insertion of alkenes into transition metal–hydrogen bonds is its reversibility, (reaction 24),

$$[(PEt_3)_2PtHCl] + C_2H_4 \underset{180\,°C,\ 1\ atm}{\overset{cyclohexane,\ 95\,°C,\ 40\ atm}{\rightleftharpoons}} [(PEt_3)_2Pt(C_2H_5)Cl] \qquad 24$$

since this provides the basis for the catalytic isomerization of alkenes (pp 57–58). In addition the insertion of alkenes into metal–hydrogen bonds is an essential part of their catalytic hydrogenation (pp 54–56), and the insertion of alkenes into metal–carbon bonds occurs during their catalytic polymerization (pp 58–61).

Acyl complexes

Metal carbonylate anion + acyl halide. The formation of an acyl complex by the reaction of a metal carbonylate anion and an acyl halide is essentially the same as on p 17 (reaction 14).

Carbonylation of a metal alkyl. Acyl complexes can often be prepared by the carbonylation of a metal–alkyl, a reaction that is often reversible (reaction 25).

$$[(PEt_3)_2Pt(CH_3)Cl] \underset{-CO,\ 140\,°C}{\overset{+CO,\ 90\,°C,\ 80\ atm}{\rightleftharpoons}} [(PEt_3)_2Pt(COCH_3)Cl] \qquad 25$$

The most thoroughly investigated carbonylation is that of methylpentacarbonylmanganese(I) where it was shown that: (*a*) If labelled carbon monoxide is introduced, it ends up not in the acyl group but *cis* to it (reaction 26).

$$[MeMn(CO)_5] + {}^{14}CO \overset{pressure}{\longrightarrow} [(MeCO)Mn(CO)_4({}^{14}CO)] \qquad 26$$

Similarly in the reverse reaction ^{14}C labelled acyl complexes do not evolve labelled carbon monoxide (reaction 27).

$$[(Me^{14}CO)Mn(CO)_5] \overset{heat}{\longrightarrow} [MeMn(CO)_4({}^{14}CO)] + CO \qquad 27$$

In view of reaction 26 it is not surprising to find that ligands other than carbon monoxide can convert manganese alkyls into manganese acyls (reaction 28).

$$[MeMn(CO)_5] + L \overset{L=NH_3,PR_3\ etc}{\longrightarrow} [(MeCO)Mn(CO)_4L] \qquad 28$$

(*b*) The products formed during the decarbonylation of [(MeCO)-Mn(CO)$_4$(^{14}CO)] indicated that of the two possible mechanisms, namely carbon monoxide insertion into a manganese–methyl bond and methyl migration, the actual mechanism involved methyl migration. (*c*) The kinetics of the reaction indicated that the first stage involved methyl migration to give a 5-coordinate acyl complex, and this was then attacked by the entering carbon monoxide (or other ligand) to give the final 6-coordinate product (reaction 29).

$$[MeMn(CO)_5] \rightleftarrows [(MeCO)Mn(CO)_4] \xrightarrow{CO} [(MeCO)Mn(CO)_5] \qquad 29$$

Carbene complexes

Carbene complexes are a relatively new class of organometallic compound and new routes for their preparation are being reported frequently.   Six of the more general methods are:

(*i*) Treatment of a metal carbonyl with a lithium aryl (reaction 30) or lithium amide (reaction 31) in the presence of trialkyloxonium fluoroborate.

$$[Cr(CO)_6] + LiC_6H_4X + Me_3O^+BF_4^- \xrightarrow{Et_2O}$$

$$\left[(CO)_5Cr\left(C\!\!\begin{array}{c} \diagup OMe \\ \diagdown C_6H_4X \end{array}\right)\right] + Me_2O + LiBF_4 \qquad 30$$

$$[Mo(CO)_6] + LiNEt_2 + Et_3O^+BF_4^- \xrightarrow{Et_2O}$$

$$\left[(CO)_5Mo\left(C\!\!\begin{array}{c} \diagup OEt \\ \diagdown NEt_2 \end{array}\right)\right] + Et_2O + LiBF_4 \qquad 31$$

(*ii*) Exchange of one group within the carbene for another (reaction 32).

$$\left[(CO)_5Cr\left(C\!\!\begin{array}{c} \diagup OMe \\ \diagdown Me \end{array}\right)\right] + NH_2Et \rightarrow \left[(CO)_5Cr\left(C\!\!\begin{array}{c} \diagup NHEt \\ \diagdown Me \end{array}\right)\right] + MeOH \qquad 32$$

(*iii*) Treatment of a metal complex with an electron-rich alkene (reaction 33).

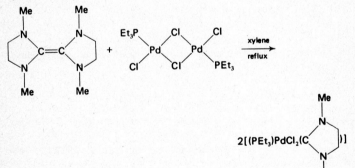

(*iv*) Treatment of a metal complex with a carbene precursor (reaction 34).

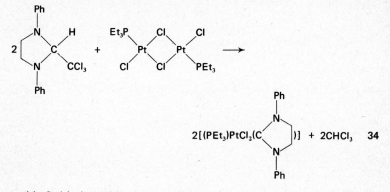

$$2\,[(PEt_3)PtCl_2(C \underset{\displaystyle N}{\overset{\displaystyle N}{}})] + 2CHCl_3 \qquad 34$$

(*v*) Oxidative-addition of an imidoyl chloride to a metal complex (reaction 35). During the reaction the oxidation state of the metal increases by 2 units.

$$n/2[Rh^I(CO)_2Cl]_2 + n\ \underset{Ph}{\overset{Cl}{\diagdown}}C{=}NMe + nHCl \rightarrow$$

$$\left[(CO)Cl_3Rh^{III}\left(C\overset{NHMe}{\underset{Ph}{\diagup}}\right)\right]_n \qquad 35$$

(*vi*) Nucleophilic attack on a coordinated isocyanide ligand (reaction 36).

$$cis\text{-}[(PEt_3)PtCl_2(CNPh)] + EtOH \rightarrow cis\left[(PEt_3)Cl_2Pt\left(C\overset{OEt}{\underset{NHPh}{\diagup}}\right)\right] \qquad 36$$

Alkene and alkyne complexes

Alkene and alkyne complexes are most commonly prepared by the displacement of other coordinated ligands by alkenes (reactions 37–39).

$$Na_2PtCl_4 + CH_2{=}CH_2 \xrightarrow{\ H_2O\ } Na[Pt(CH_2{=}CH_2)Cl_3] + NaCl \qquad 37$$

$$CH_2{=}CH_2 + [(\pi\text{-}C_5H_5)Mn(CO)_3] \xrightarrow[\text{in petroleum}]{\text{uv irradiation}}$$

$$[(\pi\text{-}C_5H_5)Mn(CO)_2(C_2H_4)] + CO \qquad 38$$

$$[Pt(PPh_3)_4] + PhC{\equiv}CPh \xrightarrow[25\,°C]{CH_2Cl_2} [(PPh_3)_2Pt(PhC{\equiv}CPh)] + 2PPh_3 \qquad 39$$

π-Allylic complexes

Three main routes have been used for the preparation of π-allylic complexes.

From allyl halides or allyl alcohols. Reaction of allyl chloride with a carbonylate anion yields a σ-allyl complex together with sodium chloride, the latter providing the driving force for the reaction (reaction 40).

$$Na[Mn(CO)_5] + ClCH_2CH{=}CH_2 \rightarrow [(CO)_5Mn(CH_2CH{=}CH_2)] + NaCl$$

40

If the product is irradiated with ultraviolet light or heated to 80 °C one carbon monoxide ligand is lost and the σ-allyl rearranges to a π-allyl (reaction 41).

$$[(CO)_5Mn(CH_2CH{=}CH_2)] \xrightarrow{\text{uv or 80 °C}} [CH{\underset{CH_2}{\overset{CH_2}{\diagup}}}Mn(CO)_4] + CO \qquad 41$$

Although no thermodynamic data are available for this reaction, a considerable part of the driving force is undoubtedly provided by the gain in entropy of the system consequent to the release of a carbon monoxide molecule.

Allyl chloride and allyl alcohol react with Na_2PdCl_4 to yield π-allylpalladium chloride in a reaction (reaction 42) that involves (*a*) the formation of an alkene complex, (*b*) nucleophilic attack on this alkene complex to yield palladium(0) (pp 49–53) and (*c*) oxidative addition of a further molecule of allyl chloride or allyl alcohol to the palladium(0) to yield $[(\pi\text{-}C_3H_5)PdCl]_2$.

$$2CH_2{=}CHCH_2X + 2Na_2PdCl_4 \xrightarrow[\text{CH}_3\text{COOH}]{\text{50\% aq.}} \left[\text{...} \right] \qquad 42$$

$$(X{=}Cl, OH)$$

From alkenes. Alkenes react with some metal salts such as palladium(II) chloride to yield olefin complexes which lose hydrogen chloride to yield π-allylic complexes either spontaneously or in the presence of added base (reaction 43).

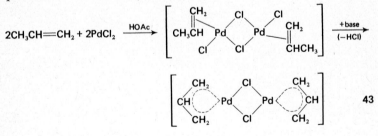

When conjugated dienes such as butadiene react with metal salts to yield π-allylic complexes, the diene gains an extra group (Cl in reaction 44); this is in direct contrast to the alkenes which must lose a group (H in reaction 43).

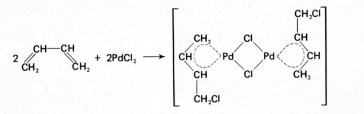

44

From allyl Grignard reagents. Although the reaction of allyl Grignard reagents does give π-allylic complexes it is only rarely used (reaction 45),

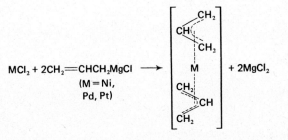

45

firstly because allyl Grignard reagents are difficult to make because of their propensity to react with further allyl chloride to yield biallyl (hexa-1,5-diene), itself a good ligand, (reaction 46), secondly because the other methods described above are experimentally simpler, and thirdly because many π-allylic complexes are susceptible to hydrolysis, which is the normal method for working up a Grignard reaction.

$$CH_2=CHCH_2MgCl + ClCH_2CH=CH_2 \rightarrow$$

$$CH_2=CH-CH_2-CH_2-CH=CH_2 + MgCl_2 \quad 46$$

Complexes of 4-carbon-bonded ligands

Direct reaction. The complexes of most 4-carbon-bonded ligands are prepared directly by displacing an existing ligand (*e.g.* CO) from the metal by direct reaction with the 4-carbon-bonded ligand (reactions 47 and 48).

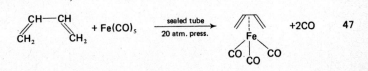

47

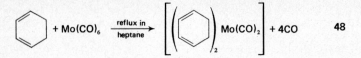

Indirect reaction. Complexes of cyclobutadiene and substituted cyclobutadienes cannot be prepared directly because the ligands do not exist in the free state. The commonest route to these compounds is by dimerization of the disubstituted alkyne in the presence of a metal complex (reaction 49), although a more specialized preparation for complexes of cyclobutadiene itself is required (reaction 50).

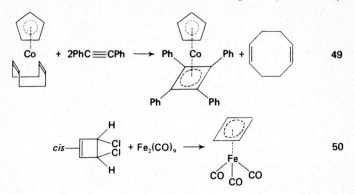

Complexes of 5-carbon-bonded ligands

There are two main methods for preparing complexes of 5-carbon-bonded ligands. In the first a metal halide is treated with sodium cyclopentadienide (prepared as in reaction 51) to yield the π-cyclopentadienyl complex and sodium halide (reaction 52).

$$(C_5H_6)_2 \xrightarrow{\text{heat}} 2C_5H_6 \xrightarrow{+2\text{Na in THF}} 2NaC_5H_5 + H_2(g) \qquad 51$$

$$NiCl_2 + 2NaC_5H_5 \xrightarrow{\text{THF}} [(\pi\text{-}C_5H_5)_2Ni] + 2NaCl \qquad 52$$

The sodium halide, of course, provides the main thermodynamic driving force for the formation of the endothermic π-cyclopentadienyl complex as can be seen from the thermodynamic data for reaction 53 which has an overall favourable enthalpy change of -483.1 kJ mol^{-1}.

$$2Na + 2C_5H_6 + FeCl_2.4H_2O \longrightarrow$$

$\Delta H_f^{\ominus}$ (kJ mol^{-1})   0 $+105.1$ -1552.0

$$[(\pi\text{-}C_5H_5)_2Fe] + 2NaCl + H_2(g) + 4H_2O$$

$+141.1$ -411.0 0 -286.0

$$\Delta H_{\text{reaction}}^{\ominus} = -483.1 \text{ kJ mol}^{-1} \qquad 53$$

As an alternative the reaction can be carried out in the presence of base (reaction 54).

$$2C_5H_6 + FeCl_2 + 2Et_2NH \rightarrow [(\pi\text{-}C_5H_5)_2Fe] + 2Et_2NH_2Cl \qquad 54$$

(Experiment 3 in Chapter 7 gives details of a laboratory preparation of ferrocene suitable for use in schools.)

The second main route to π-cyclopentadienyl complexes involves the displacement of two neutral ligands such as carbon monoxide from the metal by cyclopentadiene, followed by loss of a hydrogen atom from the ring (reaction 55).

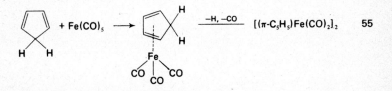

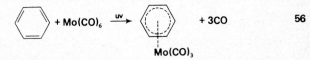

Complexes of 6-carbon-bonded ligands

Complexes of 6-carbon-bonded ligands such as benzene are usually prepared directly as in reaction 56 where a major part of the thermo-dynamic force is provided by the entropy gained by the displaced gaseous carbon monoxide ligands.

The metal atoms in π-arene complexes are generally in a low oxidation state so that it is often necessary to reduce a salt of the metal. In reaction 57 aluminium powder is used to reduce chromium(III) to chromium(I), and aluminium trichloride is used to catalyse the addition of benzene to chromium(I).

$$3CrCl_3 + 2Al + AlCl_3 + 6C_6H_6 \rightarrow 3[(\pi\text{-}C_6H_6)_2Cr]^+[AlCl_4]^- \qquad 57$$

Complexes of 7-carbon-bonded ligands

In the preparation of π-cycloheptatrienyl complexes from cyclo-heptatriene a hydrogen atom must be lost from the ligand. This either occurs spontaneously as in reaction 58 or in the presence of a hydride abstracting reagent such as the triphenylmethyl cation which converts a 6-carbon-bonded ligand into a 7-carbon-bonded ligand (reaction 59).

$$[(\pi\text{-}C_5H_5)V(CO)_4] + C_7H_8 \rightarrow [(\pi\text{-}C_5H_5)V(\pi\text{-}C_7H_7)] + 4CO \qquad 58$$

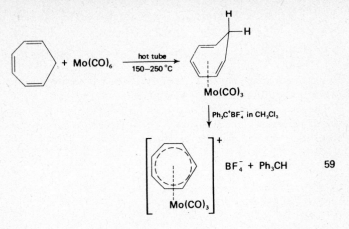

Complexes of 8-carbon-bonded ligands

Complexes of 8-carbon-bonded ligands such as $[U(C_8H_8)_2]$ are prepared by treating a metal halide with the potassium salt of cyclooctatetraene (reaction 60)

$$4K + 2C_8H_8 \xrightarrow[-30\,°C]{THF} 2(K^+)_2C_8H_8^{2-} \xrightarrow[THF\ at\ 0\,°C]{+UCl_4\ in} [U(C_8H_8)_2] + 4KCl \quad 60$$

to yield the sandwich type product together with potassium chloride, which undoubtedly provides the thermodynamic driving force for the reaction.

Low temperature condensation of high temperature species

A novel experimental technique for forming organometallic compounds that has recently been developed involves vaporizing a metal into a vessel containing an organic vapour under low pressure.   The walls of the vessel are cooled to $-196\,°C$ with liquid nitrogen and the products of the reaction condense on the cold surface.   The metal is vaporized from an alumina-coated, resistively heated, conical spiral of molybdenum wire.   The method, which is still in its infancy, has been used to prepare a number of metallocenes (reaction 61) and π-allylnickelhalides (reaction 62).

$$Fe(g) + C_5H_6(g) \xrightarrow{cocondense} [(\pi\text{-}C_5H_5)_2Fe] + H_2 \quad 61$$
$$(70\%\ yield)$$

$$Ni(g) + CH_2{=}CHCH_2Cl(g) \xrightarrow{cocondense} [(\pi\text{-}C_3H_5)NiCl]_2 \quad 62$$
$$(70\%\ yield)$$

3. Structures of Organometallic Compounds

A number of physical techniques are used in the determination of the structures of organometallic compounds. One of the most useful is undoubtedly nmr, but infrared and occasionally Raman spectroscopy are also widely used. Confirmations of structures postulated on the basis of these methods is usually obtained by x-ray diffraction techniques, although it must be remembered clearly that x-ray diffraction studies are carried out on solid crystalline samples where nmr is usually carried out in solution. There are a considerable number of compounds known which have different structures in the solid state from their structures in solution, in which case x-ray diffraction and nmr indicate different structures for the same compound.

In this chapter the structures of the main classes of organometallic compounds will be looked at briefly.

One-carbon-bonded hydrocarbon complexes

Compounds which contain alkyl ($-CH_3$), aryl ($-C_6H_5$), alkenyl ($-CH=CH_2$) and

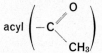

acyl

groups have essentially similar structures to the corresponding chlorides — methyl chloride, chlorobenzene, vinylchloride and acetyl chloride — with the metal occupying the same position as the chlorine atom (XII–XV). Unless the other ligands bound to the metal sterically prevent it, there is free rotation about the metal–carbon bond.

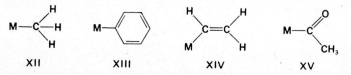

XII	XIII	XIV	XV

The hydrocarbon ligand occupies a normal ligand site on the metal. Thus tetraethyllead involves a tetrahedral arrangement of ethyl groups around a central lead atom. Trimethylboron has a

$$H_3C-B\begin{matrix}CH_3\\CH_3\end{matrix}$$

planar arrangement of three methyl groups around a central sp^2-hybridized boron atom. There is a vacant empty p-orbital perpendicular to the BC$_3$ plane which is a potential electron acceptor (Lewis acid). Accordingly B(CH$_3$)$_3$ readily forms complexes with electron donors (Lewis bases) such as ammonia (*i.e.* (CH$_3$)$_3$B← NH$_3$).*

Although alkyl groups normally bond to only one metal atom they may, on occasion, bond to more than one and, in fact, do so in the electron deficient alkyls of aluminium, beryllium, magnesium and lithium. Thus in trimethylaluminium two of the methyl groups are bound to two metal atoms whilst the other four bind to only one giving a dimeric structure

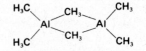

In dimethylberyllium all the methyl groups are linked to two beryllium atoms giving a linear chain structure:

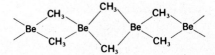

In methyllithium each methyl binds to three lithium atoms, so that the structure is built up of tetrameric Li$_4$Me$_4$ units. Since the lithium atoms carry a residual positive charge and the methyl groups residual

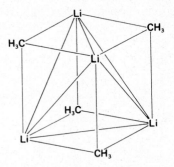

negative charges, the positive lithium atoms of one tetramer interact with the negative methyl groups of another, so that methyllithium is infusible on heating.

* Note on forming such a complex the boron rehybridizes from sp^2 in B(CH$_3$)$_3$ to sp^3 in the complex, so that in the complex the groups around the boron lie at the points of a tetrahedron.

Carbene complexes

Carbene complexes, in which a

group is bound to the metal, have a planar

unit and if there are only three other ligands bound to the metal

i.e.
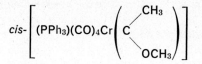

then this plane lies perpendicular to the plane containing the metal carbene carbon and three other ligands; if there are five other ligands then this plane lies at 45° to each of the planes containing the metal, carbene carbon and three other ligands (*Fig. 3*). Although there is essentially free rotation about the metal–carbon bond, subject to the steric requirements of the other ligands bound to the metal, nmr has shown that there is restricted rotation about C—O or C—N bonds in the carbene moiety. Thus for example

$$cis\text{-}\left[(PPh_3)(CO)_4Cr\left(C\begin{smallmatrix}\nearrow CH_3\\ \searrow OCH_3\end{smallmatrix}\right)\right]$$

exists in two isomeric forms (XVI and XVII) which can be distinguished at low temperature by nmr.

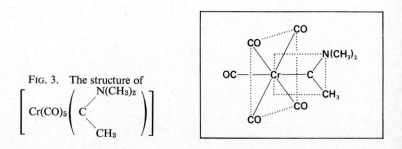

Fig. 3. The structure of
$$\left[Cr(CO)_5\left(C\begin{smallmatrix}\nearrow N(CH_3)_2\\ \searrow CH_3\end{smallmatrix}\right)\right]$$

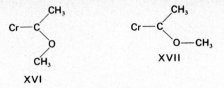

XVI   XVII

Two-carbon-bonded complexes

The essential feature about the structure of the complexes of sym-
metrical alkenes and alkynes is that the two carbon atoms adjacent to
the double-bond are equidistant from the metal (XVIII and XIX) —

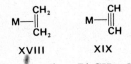

XVIII   XIX

with unsymmetrical alkenes such as PhCH=CH₂ some distortion of
this ideal geometry does occur.   The angle between the plane of the
multiple bond and the rest of the molecule depends on the number of
other ligands present.   If there are three other ligands then the
multiple bond lies approximately perpendicular to the plane containing these ligands and the metal (*Fig. 4a*), whereas if there are two
or four other ligands then the plane of the multiple bond lies close to
the trigonal plane of the molecule (*Figs 4b and 4c*).   X-ray diffraction
is generally unable to indicate the positions of hydrogen atoms in
these alkene complexes which, of course, contain strong x-ray
scattering heavy metal atoms.   However, where there are other
stronger x-ray scattering substituents present then it is found that
these substituents are bent out of the plane of the multiple-bond away
from the metal.

There is some lengthening of the multiple bond on coordination.
However, this varies from a very slight lengthening in complexes of
platinum(II) and palladium(II) to a considerable lengthening in

FIG. 4.   The structures of (*a*) [Pt(C₂H₄)Cl₃]⁻, (*b*) [(PPh₃)₂Pt(C₂H₄)] and (*c*)
[(PPh₃)₂IrH(CO){C₂(CN)₄}].

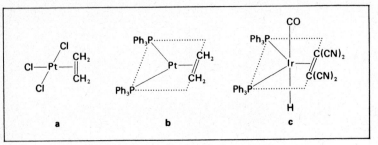

a   b   c

complexes of platinum(0), nickel(0), iron(0) and iridium(I). Nmr further indicates that the alkenes in the complexes where there is only slight lengthening of the double-bond on coordination have only low activation energies to their rotation about the metal–alkene bond axis, whereas in the complexes in which there is considerable lengthening of the double-bond, then this double-bond is essentially held rigid. These results have been summarized by suggesting that there are essentially two classes of olefin complex (S and T) whose properties are summarized in Table 3.

Table 3.   Classification of alkene complexes[a].

Class	S	T
Model complex	K[Cl—Pt(Cl)(—∥CH₂ CH₂)]	Ph₃P, Ph₃P—Pt(=CH₂ CH₂)
Coordination number	4 or 6	3 or 5
Angle between axis of double-bond and plane containing the metal atoms	Approximately 90° (in practice 77–90°)	Approximately 0° (in practice 0–24°)
Multiple bond lengthening on coordination	Slight (~2 pm)	Considerable (~15 pm)
Angle at which substituents on the multiple-bond are bent back away from the metal	~15°	~35°
Rotation about the metal–alkene bond?	Yes	No
Examples of metal ions forming complexes	Pt^{II}, Pd^{II}, Fe^{II}, Rh^{I}, Re^{I}, Mn^{I}	Pt^0, Pd^0, Fe^0, Ir^{I}, W^{I}, Mo^{I}

[a] The letters S and T are derived from the Square planar and Trigonal structures of the two model complexes.

Three-carbon-bonded complexes

π-Allylic complexes have structures in which the three carbon atoms bound to the metal are all approximately equidistant from the metal. Thus, for example, in π-allylpalladium chloride (*Fig. 5*) the palladium–carbon bond lengths are equal to within 1 pm. This is achieved by having the plane of the allylic group at about 111.5° to the plane containing the palladium and the other ligands. A similar arrangement is found in the π-allylic complexes of other metals. When the metal is bound to two different ligands, as in [(π-C₃H₅)Pd(SnCl₃)-(PPh₃)] (XX), then the asymmetry of the ligands on the right hand

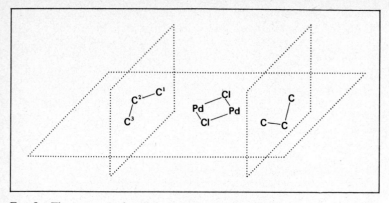

Fig. 5.   The structure of π-allylpalladium chloride. The palladium–carbon bond lengths are: Pd–C^1 212 pm, Pd–C^2 211 pm, Pd–C^3 212 pm.

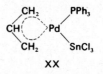

XX

side of the complex causes an asymmetry in the palladium–carbon bond lengths.   Other features of the structures of allylic complexes are:

(*i*) In the absence of asymmetry elsewhere in the complex the two carbon–carbon bond lengths in the π-allylic moiety are virtually equal.

(*ii*) Methyl substituents present on the allylic group are distorted out of the plane of the allylic group *towards* the metal — this is in contrast to alkene complexes where such substituents are bent *away* from the metal.   The allylic hydrogen atoms appear to be coplanar with the allylic group's plane, although since the principal structural technique used in this connection is x-ray diffraction, it is possible that the limited ability of this technique for detecting hydrogen atoms is inadequate to detect any non-planarity.

Complexes of four-carbon-bonded acyclic ligands such as butadiene

In the complexes of acyclic four-carbon-bonded ligands as exemplified by butadiene the butadiene group itself is *cis* and planar.   It lies above the plane of the metal atom and approximately equidistant from all four carbon atoms.   This in turn requires that in a complex such as $[(C_4H_6)Fe(CO)_3]$ the plane of the butadiene ligand is not

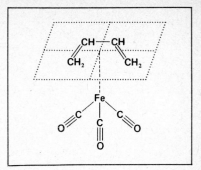

FIG. 6.   The structure of butadiene iron tricarbonyl. The Fe—$C_{butadiene}$ bond lengths are all 210 ± 4 pm.

quite parallel to that formed by the three carbon atoms of the carbonyl group (*Fig. 6*).

Complexes of cyclic 4-, 5-, 6-, 7- and 8-carbon-bonded ligands

In considering the structures of complexes containing cyclic ligands it is convenient to split the complexes up into two groups, the first containing ligands in which all the carbon atoms in the ring are bound to a single metal atom and the second complexes in which not all the carbon atoms in the ring are bound to the metal atom.

All carbon atoms bound to a single metal

The 4-, 5-, 6-, 7- and 8-carbon-bonded ligands in which all the carbon atoms are bound to a single metal atom have a number of features in common.   Thus the rings are planar, all the carbon atoms in the rings are approximately equidistant from each other and where the free ligands are available for study it is found that the C—C bond lengths in the coordinated ligands are greater than in the free ligands. All the metal–carbon bond lengths are also approximately equal. For the 5- and 6-carbon-bonded ligands, π-cyclopentadienyl and benzene, there is a low energy barrier to rotation of the ring about the metal-ring axis.   As a result inter-ring forces are important in determining the structures of metallocenes, so that in ferrocene and dibenzene chromium, with small first transition series atoms, inter-ring repulsion leads to staggered conformations (*Figs 7a* and *7b*) whereas in ruthenocene, where the metal atom is larger, an eclipsed conformation is observed (*Fig. 7c*).

It is impossible to generalize about the positions of substituents in these complexes since in tetraphenylcyclobutadiene complexes such as $[(C_4Ph_4)Fe(CO)_3]$ the phenyl groups are bent out of the plane away from the metal atom, whereas in hexamethylbenzene complexes the methyl ligands are essentially coplanar with the plane of the benzene ring.

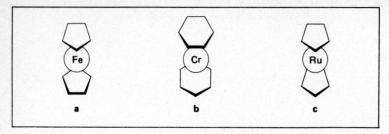

FIG. 7.   The staggered conformations of (*a*) ferrocene and (*b*) dibenzenechromium in contrast to the eclipsed conformation of ruthenocene (*c*).

Some of the ring carbons not bound to the metal

Compounds of 6-, 7- and 8-membered ring compounds in which some of the carbon atoms are not bound to the metal are known — *e.g.* π-cyclohexadienyl (XXI), π-cycloheptatrienyl (XXII) and π-cyclooctatrienyl (XXIII).   In these the bound carbon atoms lie in a plane and the metal atom lies beneath this plane approximately equidistant

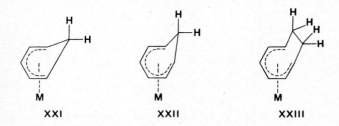

from each of the bound carbon atoms.   The unbound carbon atoms lie above the bound carbon plane away from the metal.   The carbon–carbon bond lengths are not equal but in the six-carbon-bonded ligands alternate in length as expected for double and single bonds.

Fluxional molecules

There are a number of organometallic compounds known where different physical techniques indicate different structures.   In particular x-ray diffraction, which is an essentially time-independent technique carried out on crystalline samples, often indicates a different structure to nmr, which is a time-dependent technique carried out in solution.   Further, by lowering the temperature of the nmr measurement, the spectrum sometimes changes to one that is consistent with the x-ray diffraction data.   Such a system is described as fluxional.   An example is the molecule $[(\pi\text{-}C_5H_5)Fe(CO)_2(\sigma\text{-}C_5H_5)]$, which x-ray diffraction indicates has structure XXIV.   The σ-cyclopentadienyl ligand in XXIV would be expected to exhibit three proton

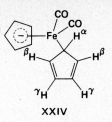

XXIV

nmr resonances due to H^α, H^β and H^γ, which are all inequivalent,
However, at room temperature these only show a single resonance,
although on cooling in carbon disulphide solution to $-80\,^\circ$C the
expected three resonances are observed.   The explanation is that at
low temperature the structure is effectively frozen into that observed
in the crystal, but as the temperature rises a series of 1,2-shifts by the
iron atom result in all the hydrogen atoms in the σ-cyclopentadienyl
ring becoming equivalent on the nmr time scale (reaction 63).

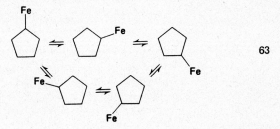

63

Fluxional behaviour has been appropriately summarized by saying
plus ça change, plus c'est la même chose (the more it changes, the more
it remains the same thing).

4. Bonding in Organometallic Compounds

As discussed in Chapter 1 organometallic compounds can be conveniently divided into ionic, covalent and electron deficient. In the ionic compounds the bonding electrons are located on the organic group which is then bound to the metal by electrostatic forces. In this chapter we shall be particularly concerned with the bonding in covalent and electron deficient organometallic compounds. In discussing the metal–ligand bond in these compounds it is convenient to consider, firstly the orbitals available for bonding on the free metal and the free organic group separately, and then to consider how these orbitals interact to bind the two parts of the organometallic molecule together.

Metal orbitals

In considering covalent organometallic compounds we shall be concerned with three types of metal:

(*i*) Non-transition metals, which have *s*- and *p*-orbitals in their valence shells;

(*ii*) Transition metals which have *s*-, *p*- and *d*-orbitals in their valence shells;

(*iii*) Lanthanide and actinide metals which have *s*-, *p*-, *d*- and *f*-valence shells.

In order to determine the number of electrons in these valence shells it is convenient to assign an oxidation state to the metal, following the rules described in Chapter 1 (pp 7–9), and to consider the organic group as uni-negatively charged if it binds to the metal with an odd number of carbon atoms (*e.g.* CH_3^-, $CH_2 \cdots CH \cdots CH_2^-$, $C_5H_5^-$), or neutral if it binds to the metal through an even number of carbon atoms ($CH_2{=}CH_2$, butadiene, benzene, cyclooctatetraene).

In a typical non-transition metal compound such as tetraethyllead these rules will result in the lead atom at the centre being stripped of its four valence shell electrons. If we consider the lead as being sp^3 hybridized then it will have four empty orbitals pointing to the apices of a regular tetrahedron. This picture of a metal atom as being stripped of all its valence shell electrons is typical of a non-transition metal. However with transition metals a different picture emerges since, although transition metals in their valence state have some empty acceptor orbitals, they also have some filled orbitals. These filled orbitals on the transition metal can also take part in bonding by donating part of their electron density back to the ligand and so giving rise to what is known as back-donation from metal to ligand. It is this back donation that provides the major difference between transition and non-transition metals and enables the former to bond to more than one carbon atom of an organic group as in alkene, π-allylic *etc* complexes. This will be considered in more detail shortly (pp 39–44). The lanthanide and actinide metals resemble the transition metals in having both empty and filled orbitals in their valence shells.

Ligand orbitals

The simplest way to determine how a ligand can bond to a metal is first of all to build up the σ-bond framework of that ligand and then see what orbitals are left over. This is illustrated schematically in Table 4. Rather than work through all the ligands in Table 4 some specific examples will be taken.

Methyl and phenyl ligands

By the rules enunciated earlier (p 8) the methyl ligand is negatively charged (*i.e.* CH_3^-) in organometallic compounds. After building up the σ-bond framework one sp^3 orbital remains and this contains two electrons which may be used to form a bond with a metal. An exactly analogous situation arises in phenyl ligands ($C_6H_5^-$) except that the orbital on the phenyl group that contains two electrons is an sp^2 orbital.

Ethene

Once the σ-bond framework of ethene has been formed each carbon atom is sp^2 hybridized and contains a p-orbital with a single electron in it. These p-orbitals interact to give two molecular orbitals known as a π-bonding molecular orbital and a π^*-antibonding molecular orbital (64).

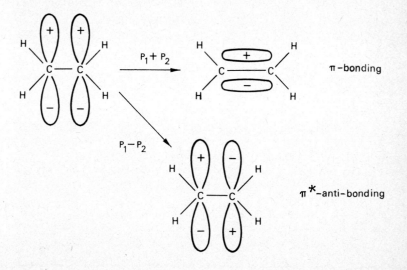

The original two electrons from the two p-orbitals now go into the lower energy molecular orbital, which is the π-bonding orbital.   Thus ethene has two molecular orbitals available for bonding to a metal, the π-bonding orbital which contains two electrons and the π^*-antibonding orbital which is of slightly higher energy and which is empty.*

Higher ligands

A careful study of Table 4 will reveal that the number of p-orbitals left over after the σ-bond framework of the organic ligand has been constructed is equal to the number of carbon atoms in the ligand, and further that there is always one p-orbital on each carbon atom.   It is a rule that when n atomic orbitals are combined to form molecular orbitals, n molecular orbitals result.   This is also apparent in Table 4.

Metal–ligand orbitals

Now that we have metal orbitals and ligand orbitals, how do these combine when we bring the organic group up to the metal to form an organometallic compound?

Metal–alkyls

Let us consider the methyl group and consider initially metal methyls in which the methyl group bonds to only one metal atom (*e.g.* BMe_3, $PbMe_4$ or $[(PPh_3)_2PtMe_2]$).   It has just been shown that a methyl group has two electrons in an sp^3 hybrid orbital.   These can be donated to a suitable empty orbital on the metal to give a metal–methyl single bond (known as a σ-bond).   An exactly analogous

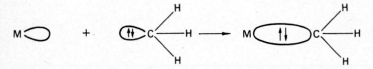

situation arises in phenyl ligands ($C_6H_5^-$) except that the orbital on the phenyl group that donates electrons to the metal is an sp^2 hybrid orbital.   In the absence of any further bonding or steric effects there will be free rotation about metal–alkyl and metal–aryl bonds.   This is a consequence of these bonds being σ-bonds, and for the purposes of this monograph σ-bonds may be defined as bonds which have no

---

* The reader will notice that in considering methyl and phenyl ligands no mention was made of anti-bonding orbitals.   This was because with these ligands the anti-bonding orbitals, which are of course present, are of much higher energy than the bonding orbitals and are consequently scarcely involved in bonding to metals.

node along the bond axis, as compared to π-bonds, for example, which have one node along the bond axis.

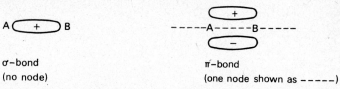

σ-bond π-bond

(no node)   (one node shown as -----)

So far we have only considered metal–methyls in which the methyl group binds to a single metal atom by a two-centre, two-electron bond. However, in electron deficient compounds one methyl group binds to two or more metal atoms (p 28). Consider trimethyl-aluminium in which aluminium is the $+3$ oxidation state. If the $3s$ and $3p$ valence orbitals of aluminium are hybridized to get four equivalent sp^3 hybrid orbitals, aluminium could be represented schematically as:

If two methyl groups are combined with the aluminium in such a way as to get two two-electron two-centre bonds we obtain:

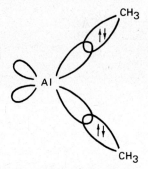

If two of these units are brought together and

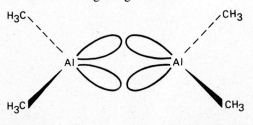

then two methyl groups put in between in such a way that their doubly occupied lobes point

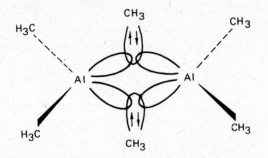

towards the midpoint between the aluminium atoms then it is possible to form two two-electron three-centre bonds. Such a structure is quite stable although, of course, the bonds between the aluminium atoms and the bridging methyl groups are not as strong as the bonds to the terminal methyl groups. The bonding in dimethylberyllium is similar to that in Al_2Me_6 except that here all the methyl groups are bridging, the structure consisting of a linear chain of sp^3 hybridized beryllium atoms with two methyl groups forming the bridge between neighbouring metal atoms. Linking up *ad infinitum* to give a linear

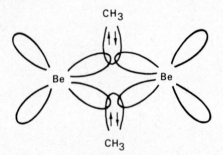

chain ensures that all the sp^3-lobes of beryllium take part in bonding. Similarly the structure of methyllithium can be understood if the valence orbitals on the lithium atoms are first hybridized to sp^3 and then all the sp^3 lobes take part in bonding by the methyl group sharing its two electrons with three lithium atoms, giving a two-electron, four-centre bond:

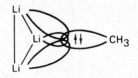

Metal–alkenes

Let us consider ethene. To a metal atom ethene will appear as

with a full π-bonding orbital (shaded) and an empty π^*-antibonding orbital (unshaded).   If now a metal atom can be found with complementary empty and full orbitals such as

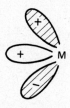

where the shaded regions are two lobes of the same orbital then bonding can occur.   If we now put on some axes

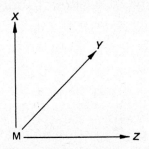

the reader will recognize the empty metal orbital as part of either the p_z or the d_{z^2} orbital and the full orbital as part of the d_{xz} orbital.   The metal and the alkene are now ideally set to combine to form two bonds each containing 2-electrons.

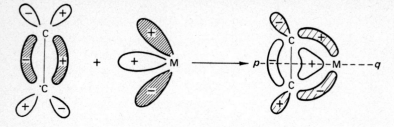

The unshaded metal–alkene orbital is known as the metal–alkene σ-bond and the shaded orbital as the π-bond (remember a σ-bond has no nodes along the molecular axis which in this case is the line pq, whereas a π-bond has just one node along the molecular axis).   It is traditional to speak of a metal–alkene bond as comprising a σ-bond formed by donation of electrons from the filled π-bonding orbital* of the alkene to an empty orbital on the metal and a π-bond formed by back-donation of electrons from a filled metal orbital of suitable symmetry to the empty π*-antibonding orbital of the alkene.   It should be noted that this bonding scheme requires the alkene to bind sideways on to the metal with two metal–carbon bonds of equal length and this is what was found in Chapter 3 when looking at the structures of alkene complexes (pp 29–31).   Furthermore, the loss of electron density from the π-bonding orbital of the alkene and the gain of electron density in the π*-antibonding orbital result in a weakening of the carbon–carbon double-bond of the alkene.

Complexes of higher ligands

Exactly the same principles as those used for ethene complexes apply when considering the complexes of higher ligands in which 3-, 4-, 5-, 6-, 7- and 8-carbon atoms are found to one metal atom.   Rather than consider all of these, let us just take one, namely cyclobutadiene and put axes on the cyclobutadiene-metal system as follows:

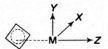

It is now apparent that the π-bonding orbital which contains two electrons (Table 4) is capable of donating these to an empty p_z or d_{z^2} orbital on the metal forming a σ-cyclobutadiene-to-metal bond, and that the two π-non-bonding orbitals on cyclobutadiene could share their single electrons with singly occupied p_x or d_{xz} and p_y or d_{yz} orbitals forming two π-bonds between cyclobutadiene and the metal.

---

* Note that this is still a π-orbital because it is on the alkene and therefore we must look for a nodal plane along the alkene axis, not along pq.

Furthermore the π^*-antibonding orbital can potentially accept electrons from a filled $d_{x^2-y^2}$ orbital on the metal.* In pairing up these metal and ligand orbitals we have simply matched up their + and − lobes to ensure that + overlaps with + and − overlaps with − (this is known as matching the symmetry of the orbitals) since bonding can only occur when the symmetries of the two interacting orbitals match. Such a bonding scheme neatly accounts for the fact that the metal bonds above the cyclobutadiene plane and is equidistant from all four carbon atoms. Furthermore we may note that cyclobutadiene itself is unknown because it has two orbitals each with one electron in it, and is thus a diradical and accordingly very reactive. However, complexes of cyclobutadiene are known, because in these the two unpaired electrons are paired by interaction with metal electrons so leading to a stable system. It is one of the triumphs of theoretical chemistry that the possibility of stable transition metal–cyclobutadiene complexes was predicted two years before they were first prepared.†

The reader may well now ask, why cannot a non-transition metal form alkene, π-allyl, cyclobutadiene *etc* complexes? This is a particularly reasonable question since in the case of cyclobutadiene complexes we have said that the p_x and p_y orbitals have the correct

Fig. 8.   Hybridization of p_y and d_{yz} orbitals.

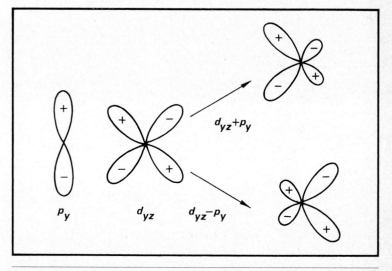

* In this case we obtain a δ-bond.   δ-bonds have two nodal planes along the bond axis (*cf.* *d*-orbitals).
† However against this must be set the theoretical chemists' prediction that the now well-established dibenzenechromium could not be prepared.

symmetry to form π-cyclobutadiene–metal bonds with the π-non-bonding orbitals of cyclobutadiene. The answer is that whilst their symmetry is correct, their spatial distribution is such that they will overlap very little with the π-non-bonding cyclobutadiene orbitals. Why then did we include them in the above discussion? The reason for this is that if a metal has both a d_{yz} and a p_y orbital and if these are hybridized together then, as shown in *Fig. 8*, we get an orbital with an even better chance of overlapping with the π-non-bonding cyclo-butadiene orbitals. Thus whilst the p_x and p_y orbitals make a contribution to the bonding in conjunction with the d-orbitals (d_{xz} and d_{yz} respectively) they are virtually useless on their own. Accordingly non-transition metals, since they lack valence shell d-orbitals, form compounds in which they bind to only one carbon atom of an organic ligand.

5. Systematic Chemistry of Organometallic Compounds

In this chapter the typical reactions of each of the classes of organometallic compounds will be described. Where possible these reactions will be illustrated by reference to commercial processes. Following the pattern of Chapters 2 and 3 compounds will be considered in order of the number of carbon atoms bound directly to the metal. However, the main emphasis will be placed on alkyl, alkene and π-allyl complexes because it is these complexes that are involved in most of the important applications of organometallic chemistry.

Hydrocarbon compounds

Hydrocarbon compounds (M—R) are not only structurally the simplest but they are also chemically the most versatile of organometallic compounds. Consequently they are formed and destroyed at the rate of thousands of tons an hour in a wide range of commercial processes. Their basic reactions are: (a) cleavage of the M—C bond, (b) insertion into the M—C bond, (c) elimination of an alkene and proton abstraction and (d) replacement of one alkyl group by another.

Cleavage of M—C bonds

Metal–alkyl and –aryl bonds are cleaved by reagents such as hydrogen, hydrogen chloride and halogens. Where possible these reactions occur by an oxidative-addition reductive-elimination mechanism. For example hydrogen chloride reacts with the 4-coordinate platinum(II) complex [(PEt$_3$)$_2$Pt(C$_6$H$_5$)$_2$] to give initially a 6-coordinate platinum(IV) complex [(PEt$_3$)$_2$Pt(C$_6$H$_5$)$_2$HCl] which subsequently eliminates a molecule of benzene to revert to a 4-coordinate platinum(II) complex [(PEt$_3$)$_2$PtCl(C$_6$H$_5$)] (reaction 65).

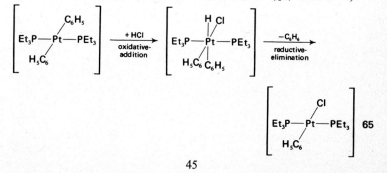

This reaction can occur again to eliminate the second phenyl group as benzene and yield $[(PEt_3)_2PtCl_2]$. The requirements for such an oxidative-addition reductive-elimination reaction are:

(a) The metal must be capable of exhibiting two oxidation states separated by two units.

(b) The coordination number of the lower oxidation state must be lower than that of the higher oxidation state (in reaction 65 it is two units lower).

Alkyl halides react similarly to hydrogen chloride yielding oxidative-addition compounds which may eliminate either the original alkyl halide or an alkane (reaction 66).

$$[(PEt_3)_2Pt(CH_3)I] + CH_3I \rightarrow [(PEt_3)_2Pt(CH_3)_2I_2]$$

80 °C
low pressure

100 °C
sealed tube

$$[(PEt_3)_2Pt(CH_3)I] + CH_3I \qquad [(PEt_3)_2PtI_2] + C_2H_6 \qquad \textbf{66}$$

Where metal–carbon bond cleavage by an oxidative-addition reductive-elimination mechanism is not possible, for example, because the metal in the metal–alkyl is either coordinatively saturated and/or in its maximum oxidation state (e.g. tin in SnR_4), such cleavage by halogens, hydrogen halides or alkyl halides occurs by an S_E2 mechanism (reaction 66a).

$$SnR_4 + XY \rightarrow \quad \underset{R}{\overset{R}{\underset{|}{Sn}}}\text{----}R\text{----}\overset{\delta+}{X}\text{----}\overset{\delta-}{Y} \rightarrow R_3SnY + RX \qquad 66a$$

Treatment of alkyl compounds with hydrogen also leads to an oxidative-addition reductive-elimination sequence of reactions which yields the corresponding alkane together with a metal–hydride (reaction 67).

$$[(PEt_3)_2Pt(CH_3)Cl] + H_2 \rightarrow [(PEt_3)_2PtHCl] + CH_4(g) \qquad 67$$

This reaction is commercially important in the hydrogenation of alkenes. Alkenes insert into metal–hydrogen bonds to form metal alkyls (pp 17–19) and these react with hydrogen to yield alkanes and the metal hydride again. It should be noted that acyl complexes react similarly with hydrogen to yield a metal–hydride and an aldehyde — a reaction that is a vital part of the commercial hydroformylation of alkenes (see reactions 102–106, pp 56–57).

Action of heat

Metal–alkyl and –aryl compounds decompose on heating. Methyl complexes appear to decompose with the formation of free radicals which, being highly reactive, combine to give the coupled products ethane and ethene (reaction 68).

$$[(PEt_3)_2Pd(CH_3)_2] \xrightarrow{100\,°C}$$
$$C_2H_6(92\%) + C_2H_4(8\%) + CH_4(trace) + 2PEt_3 + Pd^0 \qquad 68$$

With higher alkyls, however, β-hydrogen elimination to yield an alkene (below) provides a lower energy reaction pathway, so that butyl compounds, for example, yield a mixture of butene and butane by first eliminating an alkene and forming a metal-hydride (reaction 69),

$$[(PR_3)CuC_4H_9] \xrightarrow{heat} CH_3CH_2CH=CH_2 + [(PR_3)CuH] \qquad 69$$

and subsequent reaction between this metal-hydride and further butyl–metal compound (reaction 70). A β-hydrogen elimination

$$[(PR_3)CuC_4H_9] + [(PR_3)CuH] \to C_4H_{10} + 2Cu^0 + 2PR_3 \qquad 70$$

reaction of aluminium alkyls has already been mentioned as a vital step in the triethylaluminium catalysed oligomerization of ethene used in the synthesis of C_{14} alcohols (p 18).

Metal–aryl compounds are unable to undergo β-hydrogen elimination and so their decomposition yields aryl radicals, which either couple together to yield biaryls, or yield benzene derivatives by hydrogen abstraction from other groups present. It is not necessary to preform the aryl compounds in order to observe the coupling reaction. For example, treatment of palladium(II) chloride with an arylating agent yields a bis(aryl)palladium(II) compound which immediately decomposes to biaryl (reaction 71).

$$PhTl(OAc)_2 + PdCl_2 \to \text{'[PdPh}_2\text{]'} \to Ph—Ph + Pd_{metal} \qquad 71$$

Elimination of alkene

One of the most important reactions of metal–alkyls is their tendency to eliminate an alkene and form a metal–hydride with the hydrogen originally present on the β-carbon atom (reaction 72).

$$\begin{array}{c} M—CH_2 \\ \quad \\ H—CH_2 \end{array} \rightleftharpoons \begin{array}{c} M + CH_2{=}CH_2 \\ \quad \\ H \end{array} \qquad 72$$

This reaction is often reversible and thus provides a route for the synthesis of alkyls as well as their decomposition (pp 17–19). For a long time the number of transition-metal alkyl complexes was rather limited until Professor Wilkinson and his coworkers set about preventing alkene elimination from occurring. This can be achieved by one or more of the following:

(*a*) Avoid having a hydrogen atom bound to the β-carbon atom. This accounts for the much greater stability of metal–methyl compounds than the compounds of higher alkyl groups. Other groups lacking β-hydrogen atoms are $-CH_2C_6H_5$, $-CH_2SiMe_3$ and $-CH_2CMe_3$. Thus dialkylmanganese compounds containing methyl, benzyl and neopentyl ($-CH_2CMe_3$) groups are stable in tetrahydrofuran whereas those containing ethyl, propyl and butyl groups decompose fairly rapidly with elimination of the alkene.

(*b*) Have ligands other than the alkyl group present that are not readily lost from the metal thus providing a vacant coordination site that can be used in the abstraction of the β-hydrogen. Such ligands are typically tertiary phosphines (PR_3), tertiary arsines (AsR_3), carbon monoxide or π-cyclopentadienyl groups.

In the examples considered so far the β-hydrogen abstraction has been effected by the metal. However it is possible to facilitate this reaction further by adding a hydride abstractor such as the triphenyl-methyl cation (reaction 73).

$$(CO)_5Mn-CH_2CH_3 + Ph_3C^+BF_4^- \rightarrow \left[(CO)_5Mn-\overset{CH_2}{\underset{CH_2}{\|}} \right]^+ BF_4^- + Ph_3CH \qquad 73$$

Studies of this reaction with deuterium-labelled propenes have confirmed that the hydrogen abstracted was originally present on the β-carbon atom.

Insertion reactions

A number of reagents such as alkenes, carbon monoxide, sulphur dioxide and isocyanides can be inserted into metal–alkyl bonds. Alkene insertion has already been mentioned as a method of preparing metal–alkyls of both transition and non-transition metals (pp 17–19) and also forms the basis of the Ziegler–Natta process for the stereoregular polymerization of alkenes (*see* pp 59–61). Carbon monoxide insertion leads to the formation of an acyl derivative (p 19) and is an essential part of the oxo process for the hydroformylation of alkenes (*see* pp 56–57). Sulphur dioxide and alkyl isocyanides both undergo insertion reactions (reactions 74 and 75).

$$[(\pi\text{-}C_5H_5)(CO)_2FeR] + SO_2 \rightarrow [(\pi\text{-}C_5H_5)(CO)_2FeSO_2R] \qquad 74$$

$$[(PR_3)_2Pd(CH_3)I] + RNC \rightarrow \left[(PR_3)_2PdI\left(-\overset{}{\underset{NR}{\overset{\|}{C}}}-CH_3 \right) \right] \qquad 75$$

The insertion of both carbon monoxide and sulphur dioxide is often reversible and in some cases the 'deinsertion' reaction is known where the insertion is not. Thus, for example, benzenesulphinate complexes of platinum(II) eliminate sulphur dioxide to yield phenyl complexes (reaction 76),

$$[(PR_3)_2Pt(SO_2C_6H_5)Cl] \xrightarrow{\text{heat}} [(PR_3)_2Pt(C_6H_5)Cl] + SO_2 \qquad 76$$

and one of the standard reactions of acyl complexes is their loss of carbon monoxide to yield alkyl complexes (reaction 27, p 19).

Replacement reactions

The replacement of one alkyl group by another more tightly bound group is one of the standard routes for the preparation of perfluoro-alkyl complexes (reaction 77).

$$[(\text{bipyr})Pd(CH_3)_2] \xrightarrow[-CH_3I]{+C_3F_7I} [(\text{bipyr})PdCH_3(C_3F_7)] \xrightarrow[-CH_3I]{+C_3F_7I} [(\text{bipyr})Pd(C_3F_7)_2]$$

$$77$$

Such reactions probably occur, where possible, by an oxidative-addition reductive-elimination mechanism (pp 46).

Alkene compounds

Alkene compounds show a number of typical reactions such as the susceptibility of the coordinated alkene to nucleophilic attack, hydrogenation, isomerization, polymerization and disproportionation.

Nucleophilic attack

Whereas the typical reactions of free alkenes are electrophilic addition reactions (*e.g.* bromination), alkenes coordinated to metals are very susceptible to nucleophilic attack by groups such as OH^-, OAc^- and Cl^- and only in rare instances is electrophilic attack observed. Before looking in detail at specific reactions let us examine how coordination to a metal causes such a marked change in the chemical reactivity of the alkene. In electrophilic attack the electrophile approaches the alkene and accepts electron density from the π-bonding orbital of the alkene (p 37) to form a cationic intermediate (reaction 78).

$$I-H + CH_2{=}CH_2 \longrightarrow I^- + CH_3CH_2^+ \longrightarrow CH_3CH_2I \qquad 78$$

However, as noted in Chapter 4 (pp 39–42), when an alkene coordinates to a metal it donates electron density from its π-bonding orbital to suitable empty orbitals on the metal. Thus in an alkene complex there is a relatively low electron-density in the π-bonding orbital and consequently electrophilic species which, of course, seek regions of high electron density do not attack. However this region of relatively low electron density is attractive to a negatively charged nucleophile which attacks with the formation of an alkyl compound (reaction 79).

$$\overset{..}{N^-}\;\overset{CH_2}{\underset{CH_2}{\|}}\!\!-\!M \longrightarrow [N\!-\!CH_2CH_2\!-\!M]^- \qquad 79$$

In the absence of suitable stabilizing groups such an alkyl derivative is generally unstable and decomposes. In writing reaction 79 I have implied that the nucleophile is a free group that approaches the metal–alkene complex and attacks the alkene on the opposite side to the metal (*trans*-attack). Whilst in the case of palladium(II)-alkene complexes, this is true for a number of nucleophiles such as acetate ions and amines, other nucleophiles such as chloride and phenyl groups appear to coordinate to the palladium(II) before attacking the alkene (*cis*-attack). There is still some doubt about the mode of attack of some nucleophiles such as hydroxide ions, although the majority of the evidence here suggests that *cis*-attack occurs whenever possible.

Examining some specific nucleophiles in more detail:

OH^-. Hydroxide ion attack on palladium(II)–alkene complexes gives rise to carbonyl compounds (reaction 80): acetaldehyde from ethene and mainly methylketones from higher terminal alkenes (*e.g.* propene gives mainly acetone).

$$[Pd(CH_2\!=\!CH_2)Cl_2]_2 + 2OH^- \to 2CH_3CHO + 2Pd^0 + 2HCl + 2Cl^- \quad 80$$

This reaction forms the basis of the commercially important Wacker process for the one-step conversion of ethene into acetaldehyde (*see* Chapter 7 for a class experiment demonstrating this). Fortunately from a commercial point of view, it is not necessary to pre-form the palladium(II)–alkene complex, so that if ethene is bubbled into an aqueous palladium(II) salt solution, acetaldehyde and palladium metal are formed immediately (reaction 81).

$$Na_2PdCl_4 + C_2H_4 + H_2O \to CH_3CHO + Pd^0 + 2NaCl + 2HCl \qquad 81$$

Although the reaction depends on the attack of alkene by hydroxide ions, very low concentrations of the latter are necessary and so the reaction is typically carried out at about pH 4 (*i.e.* $[OH^-] \sim 10^{-10}$ M). The reaction involves:

(*a*) Coordination of the ethene to palladium with displacement of a chloride ion (reaction 82).

$$[PdCl_4]^{2-} + C_2H_4 \rightleftharpoons [Pd(C_2H_4)Cl_3]^- + Cl^- \qquad 82$$

(*b*) The ethene then helps to remove a further chloride ion and to replace it with water. Since the complexes are both square-planar with the ethene coordinated 'sideways on' this may be written schematically as in reaction 83.

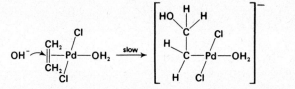

$$83$$

(*c*) The neutral aquo complex is susceptible to attack by a hydroxide ion which binds to one of the alkene carbon atoms. The other carbon atom then forms a conventional single-bond to palladium (reaction 84).

$$84$$

(*d*) This alkyl complex is unstable and decomposes. During the decomposition a hydrogen atom from the β-carbon migrates, with the aid of palladium, to the α-carbon to produce an unstable carbonium ion, which decomposes with loss of a proton to yield acetaldehyde (reaction 85).

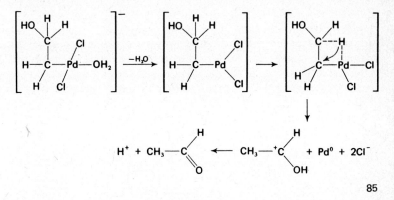

$$85$$

A reaction such as this is useless from a commercial point of view because inert palladium metal, which is very expensive, is formed in equimolar amount with the acetaldehyde. Fortunately this is not a serious problem because palladium(0) can be oxidized to palladium(II) by a number of oxidizing agents such as copper(II) chloride (reaction 86).

$$Pd^0 + 2CuCl_2 \rightarrow PdCl_2 + 2CuCl \qquad\qquad 86$$

Copper(I) chloride can in turn be reoxidized to copper(II) chloride by blowing in oxygen (reaction 87).

$$4CuCl + 4HCl + O_2 \rightarrow 4CuCl_2 + 2H_2O \qquad 87$$

Reactions 81, 86 and 87 give an overall pseudo-catalytic process for the conversion of ethene to acetaldehyde in the presence of palladium(II) salts and copper(II) chloride (reaction 88).

$$C_2H_4 + \tfrac{1}{2}O_2 \xrightarrow{\text{PdCl}_2 + \text{CuCl}_2} CH_3CHO \qquad 88$$

This process, known as the Wacker Process, is rapidly becoming the most important commercial route to acetaldehyde.

OAc⁻.   Acetate ions attack palladium(II)–olefin complexes in an analogous way to hydroxide ions, yielding vinyl acetate as the end-product (reaction 89).

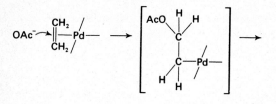

$$CH_2{=}CHOAc + H^+ + Pd^0 \qquad 89$$

Again it is not necessary to preform the palladium(II)–alkene complex so that if a mixture of ethene and air is bubbled into a solution of a palladium(II) salt and sodium acetate in acetic acid — water cannot be used as a solvent because it would lead to the formation of acetaldehyde—vinylacetate and palladium metal are formed. The palladium metal can be reoxidized as before using copper(II) salts giving an overall one-stage synthesis of vinyl acetate from ethylene (reaction 90).

$$2NaOAc + 2C_2H_4 + O_2 \xrightarrow[\text{in HOAc}]{\text{PdCl}_2/\text{CuCl}_2} 2CH_2{=}CHOAc + 2NaOH \quad 90$$

Although the Wacker process and the vinylacetate process are very similar the former is widely used commercially whereas the latter is only operated by a few companies.   The reasons are, firstly, the enormous corrosion problems encountered in the vinylacetate process which necessitate the use of a titanium- or teflon-lined reaction vessel; secondly, the vinylacetate reaction is not as clean as the acetaldehyde reaction and gives rise to by-products such as acetaldehyde, ethylidene diacetate and ethene diacetate which must be separated by careful fractional distillation; and thirdly, the acetic acid used as solvent is expensive and has a very significant influence on the economics of the reaction.

Cl^-.   Vinylchloride can be similarly prepared by passing a mixture of ethene and air through a solution of palladium(II) chloride and copper(II) chloride in an organic solvent such as benzonitrile or o-dichlorobenzene (reaction 91).

$$2CH_2{=}CH_2 + O_2 + 2Cl^- \xrightarrow{\text{PdCl}_2/\text{CuCl}_2} 2CH_2{=}CHCl + 2OH^- \qquad 91$$

However, as yet, vinylchloride is not prepared commercially in this way due to a combination of the efficiency of the alternative routes from ethene (by addition of chlorine and elimination of hydrogen chloride) and ethyne (by addition of hydrogen chloride) together with the fact that the palladium(II) catalysed reaction yields almost as much unwanted ethyl chloride as vinyl chloride.

Carbonylation of alkenes.   When a mixture of an alkene and carbon monoxide is passed into a solution of palladium(II) chloride and hydrogen chloride in an organic solvent such as benzene the alkene is converted into an acid chloride (reaction 92).

$$2CH_2{=}CHR + 2CO + PdCl_2 \xrightarrow{\text{benzene}} 2RCHClCH_2COCl + Pd^0 \qquad 92$$

By altering the solvent the acid chloride can be made to react further to yield carboxylic acids, esters, lactones or oxoacids.   The mechanism of the reaction involves nucleophilic attack on a coordinated alkene by chloride ion followed by carbon monoxide insertion into the resulting palladium–carbon bond (pp 19 and 48) (reaction 93).

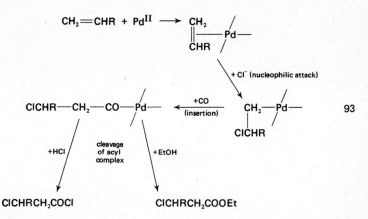

Electrophilic attack

Electrophilic attack on coordinated monoenes is unknown and is rare on coordinated polyenes. However the electrophilic attack of a triphenylmethyl cation on the bicyclo[3,2,1]octa-2,5-diene iron-tricarbonyl complex which converts it into an alkene-π-allylic complex is an example (reaction 94).

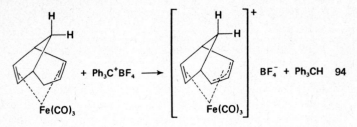

Hydrogenation

A number of transition metal salts catalyse the homogeneous hydrogenation of alkenes. The role of the metal in these reactions is threefold.

Firstly the metal provides a low energy path for cleaving the H—H bond in H_2 (bond energy 434.1 kJ mol^{-1}).

Secondly the metal coordinates the alkene, thereby weakening the bonding between the carbon atoms (pp 40–42).

Thirdly the metal provides a mechanism for transferring the two H fragments to the alkene carbon atoms so yielding an alkane.

In this section a detailed look will be taken at the mode of operation of two homogeneous hydrogenation catalysts, namely the Wilkinson catalyst ([Rh(PPh₃)₃Cl]) and pentacyanocobaltate(II) catalysts.

[$Rh(PPh_3)_3Cl$] acts as a homogeneous catalyst for the rapid hydrogenation of alkenes and alkynes at room temperature and atmospheric pressure — a school experiment based on this catalyst is described in Chapter 7. On passing hydrogen into a solution of the catalyst in benzene, an oxidative-addition reaction occurs in which the 4-coordinate rhodium(I) complex is converted to a 6-coordinate rhodium-(III) complex, which rapidly loses a triphenylphosphine ligand to give a 5-coordinate rhodium(III) complex (reaction 95).

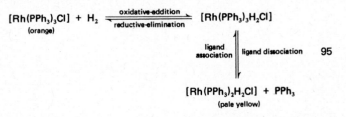

In this reaction, which is reversible, the first role of the catalyst, namely cleavage of the H—H bond is accomplished. The alkene then coordinates to the potential vacant site on the 5-coordinate dihydrido complex (reaction 96).

$$[Rh(PPh_3)_2H_2Cl] + RCH=CH_2 \rightleftharpoons [Rh(PPh_3)_2H_2Cl(RCH=CH_2)] \quad 96$$

In this step the metal has not only weakened the alkene double-bond by coordination (pp 40–42), but has also brought the alkene and the hydrogen atoms into close proximity, where they react, probably in a two-step process, to give an alkane and [Rh(PPh_3)_2Cl] which can then restart the catalytic cycle (reaction 97).

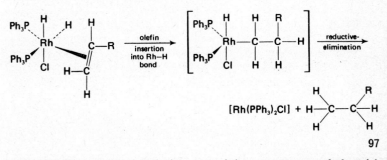

$$97$$

$[Co(CN)_5]^{3-}$.   Aqueous solutions containing pentacyanocobaltate(II) ions catalyse the homogeneous hydrogenation of conjugated dienes (to monoenes) and α,β-unsaturated aldehydes (to aldehydes), but do not reduce monoenes.   The first stage of the reaction involves cleavage of the H—H bond in hydrogen, but this time a radical mechanism (reaction 98) rather than oxidative-addition mechanism takes place.

$$2[Co(CN)_5]^{3-} + H_2 \rightleftharpoons 2[Co(CN)_5H]^{3-} \quad 98$$

The hydrido-complex may also be formed by homolytic cleavage of water (reaction 99), allowing some substrates to be reduced in the absence of molecular hydrogen.

$$2[Co(CN)_5]^{3-} + H_2O \rightleftharpoons [Co(CN)_5H]^{3-} + [Co(CN)_5OH]^{3-} \quad 99$$

The second stage involves insertion of butadiene into the Co—H bond (reaction 100).

$$[Co(CN)_5H]^{3-} + \begin{array}{c} CH-CH \\ \diagup\quad\diagdown \\ CH_2 \qquad CH_2 \end{array} \rightarrow [(CN)_5CoCH_2CH=CHCH_3]^{3-}$$

$$100$$

Although the derivative formed in reaction 100 is mainly that shown, some

$$Co-CH \begin{array}{c} \diagup CH_3 \\ \diagdown CH=CH_2 \end{array}$$

together with π-allylic products is also formed. The last stage involves the reaction of the alkyl complex with a further molecule of the hydrido-complex to yield 1-butene and the original pentacyanocobaltate(II) species (reaction 101).

$$[(CN)_5CoCH_2CH=CHCH_3]^{3-} + [(CN)_5CoH]^{3-} \rightarrow$$
$$2[Co(CN)_5]^{3-} + CH_3CH_2CH=CH_2 \quad 101$$

Other catalysts. The Wilkinson and pentacyanocobaltate(II) catalysts provide two examples of alkene hydrogenation catalysts. A number of other transition metal salts of chromium, iron, ruthenium, cobalt, rhodium, iridium, palladium and platinum also catalyse the hydrogenation of alkenes. Because the intimate mechanisms of the hydrogenations differ slightly, as well as the steric requirements of the metal and its associated ligands, some of these catalysts are very specific for particular alkenes — an effect which is often exploited, for example in the conversion of the polyene methyl linolenate to a monoene, a reaction that is an essential part of the commercial process for the conversion of soybean oil to margarine.

Hydroformylation

Hydroformylation is the addition of H_2 and CO (formally H and HCO) to a terminal alkene to yield an aldehyde which may be further reduced to an alcohol (reaction 102).

$$RCH=CH_2 + H_2 + CO \xrightarrow{\text{catalyst}} RCH_2CH_2CHO \xrightarrow{H_2} RCH_2CH_2CH_2OH \quad 102$$

The original catalyst was $HCo(CO)_4$ and the process, known as the *oxo* process, operates at about $150\,°C$ and >300 lbf in^{-2} pressure and is used to prepare about three million tons of C_7–C_9 alcohols. The reaction involves the following steps:

(*a*) Addition of the alkene to the coordinatively unsaturated $HCo(CO)_4$ followed by insertion into the Co—H bond (pp 17–19) (reaction 103).

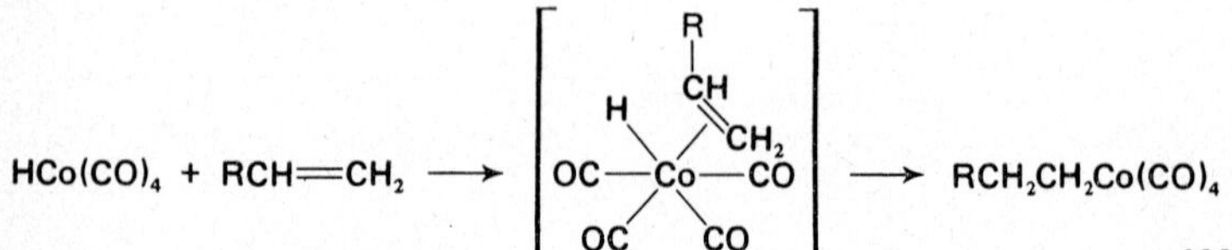

(*b*) Insertion of carbon monoxide into the Co—alkyl bond (pp 19–20) to yield an acyl complex (reaction 104).

$$RCH_2CH_2Co(CO)_4 \rightarrow RCH_2CH_2COCo(CO)_3 \quad 104$$

(*c*) Reaction of either gaseous hydrogen with the acyl complex to yield an aldehyde and $HCo(CO)_3$ (*cf.* p 46), which rapidly absorbs a further molecule of carbon monoxide to regenerate the initial catalyst $HCo(CO)_4$ (reaction 105), or of $HCo(CO)_4$ with the acyl complex to yield the aldehyde and $Co_2(CO)_8$, which in the presence of hydrogen reacts to regenerate the initial catalyst $HCo(CO)_4$ (reaction 106).

$$RCH_2CH_2-\underset{\underset{O}{\|}}{C}-Co(CO)_3 + H_2 \rightarrow$$

$$RCH_2CH_2CHO + HCo(CO)_3 \xrightarrow{+CO} HCo(CO)_4 \qquad 105$$

$$RCH_2CH_2-\underset{\underset{O}{\|}}{C}-Co(CO)_3 + HCo(CO)_4 + CO \rightarrow$$

$$RCH_2CH_2CHO + Co_2(CO)_8 \qquad 106$$

$$\underset{2HCo(CO)_4}{\overset{H_2}{\nwarrow}}$$

Since the addition of $HCo(CO)_4$ to unsymmetrical alkenes can occur in two directions and, further, since $HCo(CO)_4$ can isomerize alkenes (below), it is not surprising that mixtures of linear and branched aldehydes and ketones are often obtained from the *oxo* process. Various modified catalysts have been described for increasing the yield of linear aldehydes, for example by using tri-butylphosphine substituted cobalt carbonyls.

Isomerization

One of the major reactions of transition metal salts with alkenes is isomerization of the alkene. For this, two mechanisms have been identified. The first involves the formation of a hydrido–alkene complex which reversibly forms an alkyl complex and which gives rise to a 1,2-hydrogen shift (reaction 107).

$$RCH_2-CH{=}CH-R' + M-H \rightleftharpoons RCH_2-CH{=}CHR'$$
$$\underset{M-H}{|}$$

alkene elimination ↕ alkene insertion

$$RCH_2-\underset{\underset{M}{|}}{CH}-CH_2R' \qquad 107$$

alkene insertion ↕ alkene elimination

$$RCH{=}CHCH_2R' + M-H \rightleftharpoons RCH{=}CH-CH_2R'$$
$$\underset{M-H}{|}$$

This mechanism involves a combination of the alkene insertion (pp 17–19) and alkene elimination reactions (pp 47–48) that has already been met; the former being one of the standard routes for preparing metal–alkyls and the latter the preferred decomposition route for metal–alkyls.

The second mechanism for the isomerization of alkenes involves reaction between the alkene and a metal salt to yield a hydrido-π-allylic complex; this gives rise to 1,3-hydrogen shift (reaction 108).

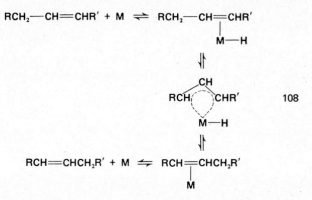

The particular mechanism operative in a given case depends principally on the metal, but also on the other ligands and the solvent present.   To be an effective alkene isomerization catalyst, a transition metal complex should either possess ligands that are readily displaced by the incoming alkene or it should be coordinatively 'unsaturated' (p 13).   The isomerization of alkenes is of major importance in the petroleum and petrochemical industries where it is normally carried out using heterogeneous transition metal catalysts (*see* Chapter 6).

Polymerization

A number of transition metal complexes catalyse the dimerization and polymerization of alkenes, but only two examples will be considered here as they illustrate the essential features of the reaction. These are the dimerization of ethene by aqueous rhodium trichloride solutions and the stereoregular polymerization of alkenes by Ziegler–Natta catalysts.

RhCl$_3$ dimerization of ethene is thought to proceed by firstly the reduction of rhodium(III) to rhodium(I) (with concomitant oxidation of ethene to acetaldehyde) and coordination of two ethene molecules to the rhodium(I) (reaction 109).

$$RhCl_3 + 3C_2H_4 + H_2O \rightarrow [(C_2H_4)_2RhCl_2]^- + CH_3CHO + HCl + H^+ \quad 109$$

This brings the two reactants close together and also by virtue of co-ordination weakens their alkene double-bonds (pp 40–42). This rhodium(I) complex then undergoes oxidative-addition of hydrogen chloride to yield an unstable 6-coordinate rhodium(III) intermediate $[(C_2H_4)_2RhHCl_3]^-$ (reaction 110).

$$[(C_2H_4)_2RhCl_2]^- + HCl \xrightleftharpoons{\text{oxidative-addition}} [(C_2H_4)_2RhHCl_3]^- \qquad 110$$

One of the ethene ligands then inserts into the Rh—H bond (pp 17–19) (reaction 111).

$$[(C_2H_4)_2RhHCl_3]^- \xrightleftharpoons{\text{alkene insertion}} [(C_2H_4)Rh(C_2H_5)Cl_3]^-$$
$$111$$

Insertion of the remaining ethene molecule into the Rh–alkyl bond followed by reductive-elimination of hydrogen chloride yields 1-butene and a rhodium(I) complex that is ready to restart the catalytic cycle by picking up two further molecules of ethene (reaction 112).

$$[(C_2H_4)Rh(C_2H_5)Cl_3]^- \xrightarrow[\text{insertion}]{\text{alkene}} [Rh(CH_2CH_2C_2H_5)Cl_3]^-$$

$$\downarrow \text{reductive-elimination} \atop \text{of HCl}$$

$$[RhCl_2]^- + C_2H_5CH{=}CH_2 + HCl \qquad 112$$

Ziegler–Natta polymerization of alkenes. The Ziegler–Natta catalysis of alkene polymerization which operates at modest temperatures and pressures, leads to the formation of stereoregular polymers. This is in direct contrast to the ICI high temperature, high pressure polymerization of ethene. The advantages of stereoregularity are that it conveys greater mechanical strength to the polymer enabling it to be used in making string and rope as well as more rigid articles such as coal hods and buckets. The high temperature, high pressure polymerized material is softer and used, for example, in polythene bags.

The basic Ziegler–Natta catalyst is prepared by treating $TiCl_4$ or $TiCl_3$ with triethylaluminium. If $TiCl_4$ is used the first reaction involves reduction of titanium(IV) to titanium(III). The catalyst is a solid heterogeneous catalyst and to understand how it works we must look at the crystal structure of $TiCl_3$. In solid $TiCl_3$ a titanium atom in the centre of the crystal is surrounded octahedrally by six chlorine atoms,

Each chlorine atom is bound to a further titanium atom or otherwise the formula would be $TiCl_6$ and not $TiCl_3$!   However if such a structure is carried on indefinitely it is apparent that, in order to make the crystal electrically neutral (*i.e.* have the formula $TiCl_3$), some of the titanium atoms at the surface will only have 5 chlorine atoms around them and so possess a vacant site ($\square$).

On treating such a crystal of titanium trichloride with triethyl-aluminium this vacant site is alkylated (reaction 113; where Cl* indicates Cl bound to a second Ti and unstarred Cl atoms lie on the surface of the crystal),

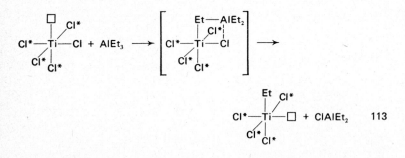

and furthermore the vacancy in the coordination shell of the titanium atom is still present although in a fresh position.   This vacant site is potentially capable of coordinating an alkene such as propene (reaction 114),

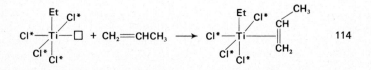

and, as found on many occasions in this chapter, when an alkene and an alkyl group are bound to adjacent sites on a metal atom they react to give what is known as an insertion product, and in so doing regenerate the vacant site enabling the process to be repeated (reaction 115).

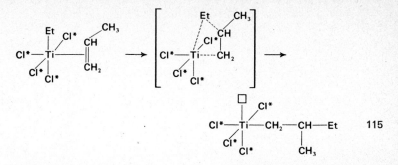

115

This yields a linear polymer, which according to the precise conditions and stereochemistry of the coordination site on the catalyst surface may be isotactic, syndiotactic or some random mixture of the two (atactic).

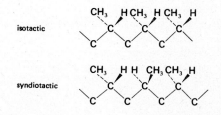

Other transition metals close to titanium in the Periodic Table act similarly. Vanadyl halides, for example, are used in the co-polymerization of styrene, butadiene and either dicyclopentadiene or 1,4-hexadiene to give synthetic rubbers.

Disproportionation

A remarkable reaction of alkenes that is heterogeneously catalysed by WO_3 at 150–500 °C and homogeneously catalysed by a mixture of WCl_6 and $EtAlCl_2$ in ethanol, is known as an alkene disproportionation (reaction 116).

$$
\begin{array}{c}
CH_2 = CHR \\
+ \\
CH_2 = CHR
\end{array}
\quad \rightleftharpoons \quad
\begin{array}{c}
CH_2 \\
\parallel \\
CH_2
\end{array}
\quad + \quad
\begin{array}{c}
CHR \\
\parallel \\
CHR
\end{array}
\qquad 116
$$

The role of the catalyst appears to be to provide two adjacent co-ordination sites for the alkene molecules which may then react together through a metallo-cyclopentane-like intermediate (reaction 117).

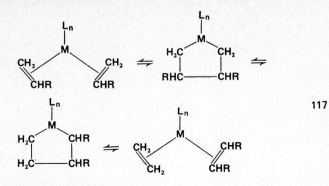

117

Alkyne compounds

The chemistry of alkyne compounds of transition metals is extremely complex and yields a great many of the more exotic organometallic compounds. The typical reactions of alkyne complexes of transition metals are hydrogenation, which is similar to that for alkenes and will not be discussed further, carbonylation and oligomerization, both of which were exploited by Reppe in his detailed investigation of ethyne chemistry in Germany before and during the Second World War.

Carbonylation

The carbonylation of ethyne by nickel carbonyl in the presence of a proton source HY yields a number of valuable organic raw materials (reaction 118).

$$HC \equiv CH + HY + CO \xrightarrow{\text{Ni(CO)}_4} CH_2 = CH - COY$$
118

$Y = OH$,   acrylic acid

$Y = OR$,   acrylic ester

$Y = NR_2$,   *N,N*-dialkylacrylamide

$Y = RCOO$, mixed anhydride

Oligomerization

Alkynes readily oligomerize in the presence of transition metals to give a remarkable range of products.

Dimerization is one of the standard routes for preparing substituted cyclobutadiene derivatives as in reaction 49 (p 24). In addition to simple dimerizations, dimerization with the incorporation of all or part of one of the ligands present on the metal are known (reactions 119, 120 and 121).

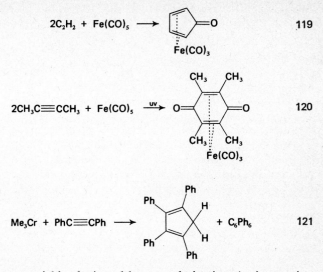

$$2C_2H_2 + Fe(CO)_5 \longrightarrow \text{(cyclopentadienone)Fe(CO)}_3 \qquad 119$$

$$2CH_3C{\equiv}CCH_3 + Fe(CO)_5 \xrightarrow{uv} \text{(tetramethylquinone)Fe(CO)}_3 \qquad 120$$

$$Me_3Cr + PhC{\equiv}CPh \longrightarrow \text{(pentaphenylcyclopentadiene)} + C_6Ph_6 \qquad 121$$

Trimerization to yield substituted benzene derivatives (as in reaction 121) can be catalysed by many transition metals. Often as in reaction 122 the products are readily explained in terms of head-to-tail polymerization but occasionally less readily explained trimers are found amongst the products (*e.g.* product A in reaction 123).

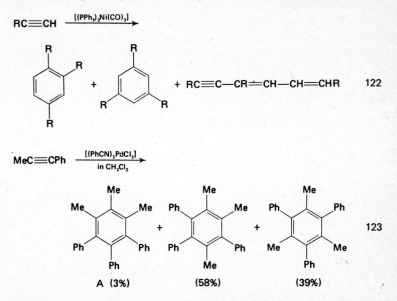

$$RC{\equiv}CH \xrightarrow{[(PPh_3)_2Ni(CO)_2]}$$

$$\text{(1,2,4-triR-benzene)} + \text{(1,3,5-triR-benzene)} + RC{\equiv}C{-}CR{=}CH{-}CH{=}CHR \qquad 122$$

$$MeC{\equiv}CPh \xrightarrow[\text{in } CH_2Cl_2]{[(PhCN)_2PdCl_2]}$$

A (3%)   (58%)   (39%)

Tetramerization can be effected by such nickel complexes as Ni(CN$_2$),

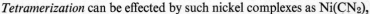

although it is generally accompanied by the formation of lower oligomers (reaction 124).

$$CH\equiv CH \xrightarrow{Ni(CN)_2}$$ 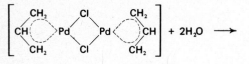 124

π-Allylic compounds

The typical reactions of π-allylic compounds are their oxidation, hydrolysis, conversion of π-allylic into σ-allylic ($M-CH_2CH=CH_2$) and their ability to insert alkenes, dienes and carbon monoxide into the metal-π-allyl bond. Nickel π-allyl complexes in particular have been shown to be of great value for catalysing the oligomerization of butadiene to give cyclic products.

Oxidation

π-Allylic complexes vary markedly in their stability to air; the palladium(II) complexes being reasonably stable at room temperature and hence, of course, widely investigated. By contrast bis(π-allyl)nickel ($[(\pi\text{-}C_3H_5)_2Ni]$) is spontaneously oxidized by air to nickel oxides.

Hydrolysis

Most π-allylic complexes are sensitive to water although the palladium(II) complexes require alkali in order to hydrolyse them. The hydrolysis of π-allylpalladium chloride is complex and yields palladium metal and a mixture of allyl alcohol, propene, acrolein and acetone. The first stage is probably nucleophilic attack upon the coordinated π-allylic group by water (cf. alkenes pp 49–50), to yield allyl alcohol and palladium metal (reaction 125).

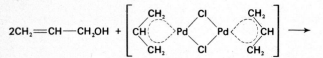

$$2CH_2=CH-CH_2OH + 2Pd^0 + 2HCl \qquad 125$$

The acrolein and propene are probably formed by reaction of the allyl alcohol formed in reaction 125 with further π-allylpalladium chloride (reaction 126).

$$2CH_2=CH-CH_2OH +$$

$$2CH_2=CH-CHO + 2CH_2=CH-CH_3 + 2Pd^0 + 2HCl \qquad 126$$

The acetone is probably formed by oxidative hydrolysis of the propene catalysed by palladium(II) (pp 49–52).

Conversion of π-allyl to σ-allyl

A π-allylic ligand can be considered to occupy two coordination sites around a metal and, accordingly, if a strongly coordinating ligand is added, the π-allylic ligand may be displaced from one of these sites to give a σ-allylic complex (reaction 127).

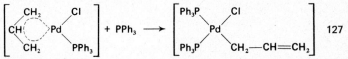

This type of reaction is a vital step in many catalytic systems involving π-allylic complexes. In this example an excess of triphenylphosphine would displace the allylic ligand altogether.

Insertion reactions

A number of organic groups can be inserted into metal-π-allyl bonds, and these insertion reactions are of great value to the synthetic organic chemist. Some, such as the ethene insertion, are the basis of industrial processes.

Ethene. The insertion of ethene into a nickel-π-allyl bond forms the basis of the du Pont route to 1,4-hexadiene which is an essential component of ternary rubbers. In this process a nickel(0) catalyst [Ni{P(OEt)$_3$}$_4$], is treated with ethene and butadiene in acid solution. The first stage involves the formation of a π-allylic complex from the nickel and butadiene (reaction 128 where L = P(OEt)$_3$ (*cf.* reaction 44, p 23)).

$$NiL_4 + H^+ + \overset{CH-CH}{\underset{CH_2 \qquad CH_2}{\diagdown\diagup}} \longrightarrow \left[\begin{array}{c} CH_3 \\ CH \\ CH \diagup\!\!\!\searrow NiL_3 \\ CH_2 \end{array}\right]^+ + L \qquad 128$$

This π-allylic complex then loses a further molecule of P(OEt)$_3$ and coordinates a molecule of ethene which can then insert into the π-allyl–nickel bond to give one of two intermediates, which then react with further P(OEt)$_3$ to liberate the organic component and yield a nickel complex that can restart the cycle by reaction with more butadiene (reaction 129).

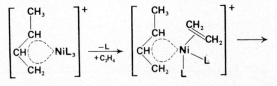

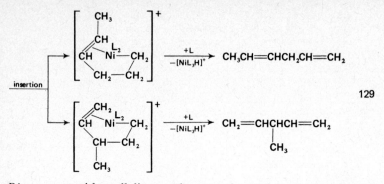

129

Dienes react with π-allylic complexes to give an insertion product in which the diene forms the π-allylic group and the original π-allylic group has become the side chain (reaction 130 where L—L = acetylacetonate).

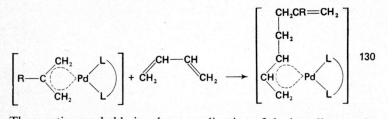

130

The reaction probably involves coordination of the butadiene to the metal and simultaneous conversion of the π-allylic group to a σ-allyl (*cf.* reaction 127) as a first step, so that the insertion reaction is strictly an insertion of the diene into a metal–alkyl bond (pp 17–19 and 48).

Carbon monoxide. The insertion of carbon monoxide into π-allyl metal bonds is of considerable synthetic use in producing unsaturated acid chlorides, unsaturated esters and so on (reaction 131).

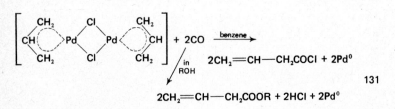

131

As with the dienes the reaction probably involves coordination of the carbon monoxide to the palladium and conversion of the π-allylic complex to a σ-allylic complex. The great synthetic utility of the reaction lies, firstly, in that it is not necessary to prepare the π-allylic complex first, since the reaction can be carried out merely in the presence of palladium(II) chloride; and secondly, a wide range of allylic compounds can be so carbonylated (reaction 132).

$$CH_2=CH-CH_2-Y + CO \xrightarrow[\text{catalyst}]{\text{PdCl}_2} CH_2=CH-CH_2-COY \qquad 132$$

Y = Cl,	yields an acid chloride
Y = OR,	yields an ester
Y = OCOCH₃,	yields a mixed anhydride

Catalysis of butadiene oligomerization

A very important aspect of π-allyl chemistry is the ability of π-allylic complexes, such as those of nickel, palladium and chromium to catalyse the dimerization and trimerization of butadiene. Thus treatment of bis(π-allyl)nickel ([Ni(π-C₃H₅)₂]) with butadiene yields a volatile blood red crystalline compound, cyclododecatrienenickel which rapidly catalyses the cyclic trimerization of butadiene (reaction 133).

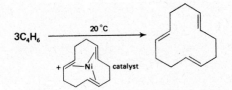

If this catalyst is treated with a molecule that is not displaced by butadiene (*e.g.* triphenylphosphine) then one of the nickel co-ordination sites becomes blocked and only dimerization of butadiene to 1,5-cyclooctadiene occurs. A whole range of cyclic compounds

1,5-cyclooctadiene

can be obtained by altering the conditions or adding other compounds such as carbon monoxide or alkenes which become incorporated in the product.

Butadiene compounds

Butadiene complexes of transition metals in which the butadiene is bound to the metal with all four carbon atoms are relatively inert, so that the butadiene group resists both hydrogenation and the Diels–Alder reaction with olefins. Indeed a great deal of chemistry can be carried out on the side chains, when these are present, without disruption of the diene–metal bond. Diene ligands can however be liberated from their complexes either by displacement with strongly coordinating ligands such as triphenylphosphine or by oxidation of the metal (*e.g.* with iron(III) or cerium(IV) salts), which, as in $[(C_4H_6)Fe(CO)_3]$, must be in a low oxidation state if it is to bind to all four of the butadiene carbon atoms.

It should be noted that many of the reactions of butadiene that are catalysed by transition metals involve butadiene coordinated to the metal by 2- or 3-carbon atoms (*i.e.* as an alkene or π-allyl ligand), in which case the butadiene moiety is far more reactive being susceptible to the typical reactions of alkene and π-allyl complexes described above.

Cyclobutadiene compounds

The reactions of cyclobutadiene complexes may be divided into two groups, reactions resulting from the aromatic properties of the cyclobutadiene ring, which are considered with those of the other cyclic complexes below, and reactions in which the cyclobutadiene is released. As already noted in Chapter 4 (pp 42–43) cyclobutadiene itself is unknown since if formed it would exist as a very reactive diradical. Thus the final products of the reactions in which cyclobutadiene is displaced from the metal are those expected from the diradicals XXV or XXVI (see reactions 134 and 135–139 in Scheme 2).

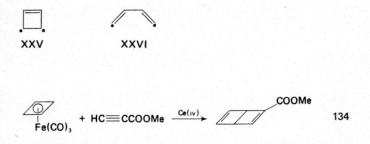

π-Cyclopentadienyl compounds

π-Cyclopentadienyl–metal complexes are fairly stable to heat and many, such as ferrocene (*see* Chapter 7, p 91), can be sublimed without decomposition. Their stability to oxidation by air varies dramatically from ferrocene which at at least 100 °C is completely inert to air, to chromocene ($[Cr(\pi-C_5H_5)_2]$) which is pyrophoric. A number of metallocenes ($[M(\pi-C_5H_5)_2]$) can be oxidized, using suitable oxidants, to the metallocenium cation $[M(\pi-C_5H_5)_2]^+$. Metallocenes are stable to water and, in contrast to free cyclopentadiene, they are very resistant to hydrogenation and do not undergo the Diels–Alder reaction, which is characteristic of uncoordinated conjugated dienes.   Those metallocenes that can be handled readily, such as ferrocene, have an extensive organic chemistry that is typical of an aromatic organic compound.

Aromatic properties of carbocyclic organometallic compounds

The organometallic complexes of cyclic ligands in which all the carbon atoms of the ligand are bound directly to the metal exhibit reactions that are typical of aromatic compounds.   Of these the first to be discovered was acetylation under Friedel–Crafts conditions

Scheme 2.

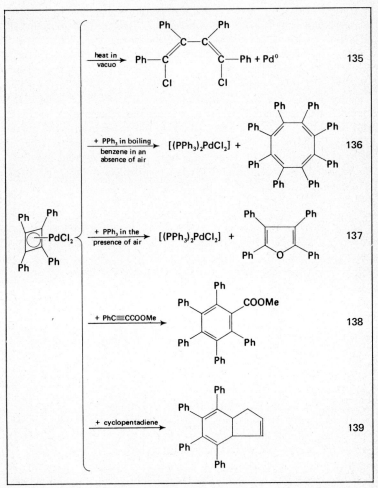

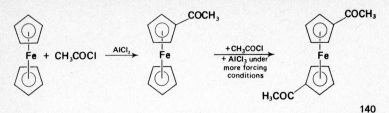

140

(reaction 140; *see* Chapter 7, p 96 for a class experiment based on this).

This reaction occurs more rapidly with ferrocene than with benzene, which, since the reaction involves electrophilic attack by the CH_3CO^+ cation, indicates the greater availability of negative charge on the π-cyclopentadienyl rings than on benzene.   Indeed the availability of electron density on the ring and hence the rate of the Friedel–Crafts reaction decreases as the size of the ring increases across the series

$$\pi\text{-}C_4H_4 > \pi\text{-}C_5H_5 > \text{benzene} > \pi\text{-}C_6H_6 > \pi\text{-}C_7H_7 > \pi\text{-}C_8H_8.$$

The role of the metal in electrophilic attack on these cyclic complexes is two-fold, firstly it initially binds the attacking electrophile and secondly, in a complex with two or more rings, the metal transmits electronic effects from one ring to another so that a deactivating substituent deactivates both rings, the second less than the first.   Thus the intimate mechanism of electrophilic attack on ferrocene is as shown in reaction 141,

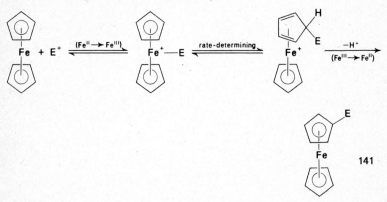

141

in which the electrophile first coordinates to the iron, oxidizing it from iron(II) to iron(III) before being transferred to the π-cyclopentadienyl ring in the rate-determining step.   The final step involves expulsion of a proton with concomitant reduction of iron(III) back to iron(II).

Other electrophilic reactions that these cyclic organometallic complexes undergo include lithiation, nitration, halogenation and cyanation all of which are, of course, typical aromatic reactions. There is thus a very wide scope to the organic chemistry of these cyclic complexes which is only significantly limited in specific cases, either by the fact that a particular complex may be unable to withstand the conditions of the reaction, or that during the reaction the metal may become oxidized which clearly inhibits electrophilic attack. In the case of ferrocene, however, its stability and the diversity of its reactions have permitted a vast range of derivatives to be prepared, although no major applications for them have yet been found.

6. Homogeneous and Heterogeneous Catalysis

During Chapter 5 both the homogeneous and heterogeneous catalysis of reactions has been referred to. In fact many reactions involving organometallic species are catalysed by both heterogeneous and homogeneous catalysts and accordingly it is worthwhile comparing the two approaches briefly.

Heterogeneous catalysts are often, but not always, transition metals finely dispersed on some inert support and in this form the metal is in the zero oxidation state — there are, of course, exceptions to this as in the Ziegler–Natta catalysts (pp 59–61). By contrast the metals in homogeneous catalysts are very often in positive oxidation states, although a number of zerovalent catalysts are known. In general, it would appear that the modes of operation of homogeneous and heterogeneous catalysts are similar in that both use the essential 'building-block' reactions discussed in Chapter 5; that is, co-ordination of the substrate, insertion, oxidative-addition, reductive-elimination and so on. However, there are clearly differences in detail due to the very different environments of a metal atom in a bulk metal on the one hand and in a complex on the other. Let us consider this by looking at some of the advantages and disadvantages of homogeneous and heterogeneous catalysts.

Advantages of homogeneous catalysts

Efficiency. In a heterogeneous system, the catalytic reaction must necessarily take place on the surface of the catalyst whereas, by contrast, all the metal atoms in a homogeneous catalyst are potentially available as catalytic centres. Thus homogeneous processes are potentially more efficient in terms of the amount of catalyst needed to catalyse a given amount of reaction.

Reproducibility. Homogeneous catalysts have the advantage over heterogeneous catalysts of being totally reproducible because they have a definite stoichiometry and structure.

Specificity. A given homogeneous catalyst has only one type of active site and therefore will often be more specific than a hetero-geneous catalyst where several types of active site may be present in the form of metal atoms in different surface defects. Furthermore, the specificity of a homogeneous catalyst can often be modified by altering the other ligands present in such a way as to alter either the electronic nature or the steric requirements of the site.

Controllability.   Closely related to specificity is the fact that, because a homogeneous catalyst has a definite structure, it is much easier to modify it in order to control the reaction.   An example of this is illustrated in Chapter 7 (*see* pp 74–82) where modification of [Rh(PPh$_3$)$_3$Cl] to [Rh(PPh$_3$)$_3$Br] causes a significant increase in the rate of hydrogenation of alkenes.

Advantages of heterogeneous catalysts

Separation of the catalyst.   The major disadvantage of homogeneous catalysts is the problem of separating the very expensive catalyst from the products at the end of the reaction.   With heterogeneous catalysts this can be achieved by some kind of coarse filtration, whereas with homogeneous catalysts a very efficient distillation is usually required.   Distillation is inevitably an endothermic process and is therefore expensive and unless it is very efficient it will result in small catalyst losses which, with expensive catalysts, can make a process uneconomic.   Sometimes, as in the preparation of margarine (*see* Chapter 7, pp 82–84), distillation is impossible and so a homogeneous catalyst, although in other ways excellent, is rendered useless.

Thermal stability.   The thermal stability of heterogeneous catalysts, particularly pure metals, is often much higher than that of homogeneous catalysts, and since the rate of many organometallic reactions increases with temperature, a high operating temperature may be an advantage.   It should be noted that high temperatures are not always ideal because some reactions involve a pre-equilibrium step which may be disfavoured by increasing the temperature.

Solvent.   Whereas the range of suitable solvents for a homogeneous catalyst is often limited, this clearly presents no problem for a heterogeneous catalyst.

One possible way, that is currently being investigated, of combining the important advantages of homogeneous catalysts with the ease of separation of a heterogeneous catalyst is first of all to develop the optimum homogeneous catalyst for the process under consideration and then to attempt to incorporate this in a resin.   For example, a homogeneous catalyst such as [(PR$_3$)$_2$MCl$_2$] might be incorporated into a polystyrylphosphine resin in such a way that one or both of the PR$_3$ groups was replaced by the phosphine of the resin.   A few examples of resin-based catalysts have been described and patented but the field is still in its infancy.   While resin-based catalysts very efficiently combine the advantages of both homogeneous and heterogeneous catalysts, it must be emphasized that they do bring their own disadvantages, the most prominent of which is the problem of diffusion of the reactants into, and the products out of, the resin.

7. School Experiments in Organometallic Chemistry

When I started to plan this monograph it was very apparent that there were few, if any, organometallic experiments suitable for use in schools. Accordingly with the aid of four very enthusiastic school teachers, who are acknowledged under the appropriate experiment, I have developed some suitable experiments. It must be emphasized that throughout we have borne in mind two factors, namely cost and availability of the equipment in a typical school. In order to minimize the former and comply with the latter we have sacrificed the yields of a number of the products described. In other words a better, but less practical, as far as schools are concerned, technique is available for some of these preparations.

The experiments described are firstly two catalytic processes, namely the Wilkinson catalyst for the reduction of alkenes (pp 54–55) and the Wacker process for the oxidation of alkenes (pp 50–52) and secondly three preparative experiments, namely the preparation of ferrocene (pp 24–25), the preparation of tetraphenyllead (pp 16–17) and the preparation of triphenylphosphine and some of its complexes.

Whenever a student comes across an unfamiliar compound we suggest that he should first consult a book such as *Hazards in the chemical laboratory* (ed. by G. D. Muir, published by The Chemical Society).

Experiment I. The homogeneous hydrogenation of alkenes using the Wilkinson catalyst [Rh(PPh₃)₃Cl]

This experiment was developed in association with Dr R. P. Smith of Richard Taunton College, Southampton.

Introduction

In this experiment the student prepares the catalyst $[Rh(PPh_3)_3Cl]$ which is fairly stable as a solid but sensitive to air in solution. He then goes on to use $[Rh(PPh_3)_3Cl]$ to homogeneously hydrogenate a simple alkene in an experiment in which he can readily study the rate of hydrogenation and also measure the total hydrogen uptake with considerable accuracy — to within 2 per cent. If a second student prepares the corresponding bromo-complex $[Rh(PPh_3)_3Br]$ they can compare the effectiveness of the chloro- and bromo-complexes and demonstrate that a very minor, but nevertheless reproducible, change in the catalyst has caused a significant change in the rate of hydrogenation (p 73). The experiment can further be used to determine

74

the molecular weight of an unknown terminal alkene and hence to identify it. Lastly by comparing the Wilkinson homogeneous catalyst with the Raney nickel heterogeneous catalyst it is possible to demonstrate the major drawback of the homogeneous catalyst, namely its separation at the end of the reaction. The chemistry of the hydrogenation is described in Chapter 5 (pp 54–55).

Cost and availability of materials

Rhodium trichloride ($RhCl_3 . 3H_2O$) is available from two main suppliers. Engelhard (Valley Road, Cinderford, Gloucestershire GL14 2PB) supply it at £1.52 per gm* (exclusive of VAT) with a minimum order of four grammes. Johnson Matthey (74 Hatton Garden, London EC1P 1AE) supply it at £2.29 per gm* or £8.95 per five grammes* (both exclusive of VAT) with a minimum order of £5. BDH, Koch-Light and other companies also sell rhodium trichloride, but as wholesalers their prices are inevitably more expensive per gramme, although they are prepared to sell one gramme amounts. Each student requires 0.2 g of rhodium trichloride which will yield about 0.6 g of catalyst, which is sufficient for at least six separate hydrogenations.

Triphenylphosphine is available from BDH and Koch-Light and costs* about £1.50 per 100 g and each student requires about five grammes. Alternatively it may be prepared as described in Experiment V (p 100). All the other chemicals should be readily available in a school laboratory. The only non-standard piece of equipment that we used was a 'suba-seal' to enable us to inject a syringe needle into a flask with standard quickfit sockets. Suba-seals for standard quickfit sockets can be obtained from Gallenkamp at a cost* of 31p per packet for B14 (note, this is size '25').

Preparation of the Wilkinson catalyst ([Rh(PPh₃)₃Cl])

Requirements

    0.2 g Rhodium trichloride ($RhCl_3 . 3H_2O$)
    5 g Triphenylphosphine
    95 per cent ethanol
    Diethyl ether
    Nitrogen (either white-spot or normal)

Method

Dissolve rhodium trichloride (0.20 g) with gentle warming in 95 per cent ethanol (15 cm³) in a two-necked 50 cm³ pear-shaped flask (B14) provided with both a nitrogen inlet, which passes to the bottom of the

---

* Note, all costs are approximate (for early 1974) and are given purely to give an idea of the overall costs. They are likely to change at any time.

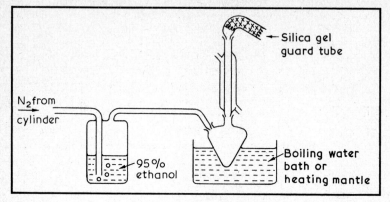

FIG. 9. Apparatus for the preparation of [Rh(PPh₃)₃Cl].

flask, and a reflux condenser protected by a silica gel guard-tube (*Fig. 9*). Pass a stream of nitrogen, presaturated with ethanol, through the solution for at least 15 minutes before adding any further reagents. Dissolve 1.2 g of triphenylphosphine* in 35 cm³ of hot deoxygenated 95 per cent ethanol (deoxygenation can be effected by bubbling nitrogen through the ethanol for 10 minutes).

Bring the rhodium trichloride solution to the boil, turn the nitrogen flow up to a brisk rate, pour the triphenylphosphine solution slowly down the reflux condenser and reflux the solution vigorously under a mild flow of nitrogen. After a few minutes an orange-brown precipitate forms, which on further vigorous refluxing darkens to a deep purple. Continue refluxing for a total of four hours (*see* note 3), before filtering the hot solution by suction, washing it three times with 5 cm³ portions of diethyl ether and sucking it dry at the pump. After storing in a vacuum desiccator overnight we obtained 0.60 g of [Rh(PPh₃)₃Cl], a yield of 80 per cent based on rhodium. The material should be stored in a polythene capped bottle and over a few days does not appear to be harmed by air, although for longer storage periods we recommend flushing the bottle with nitrogen.

Notes

1. It is strongly recommended that these preparative details be adhered to closely, for example when we added the rhodium tri-chloride solution to the triphenylphosphine (*i.e.* the opposite way to that described here) we got an orange precipitate which analysed correctly for [Rh(PPh₃)₃Cl], but which did not catalyse the hydro-

---

* Triphenylphosphine should be freshly recrystallized from 95 per cent ethanol (5 g of triphenylphosphine require 20 cm³ of ethanol), washed with diethyl ether and sucked dry in a Buchner funnel for a couple of minutes.

genation of alkenes. This orange precipitate incidentally could be converted to catalytic [Rh(PPh₃)₃Cl] by boiling it with excess triphenylphosphine.

2. The scale of the reaction can readily be increased if desired. However, the present scale yields sufficient material for at least six hydrogenations and is not unreasonably expensive.

3. If a very hot heating mantle is used instead of the vigorously boiling water bath the reaction is complete in about 30 minutes.

Chemistry

On refluxing rhodium trichloride with triphenylphosphine, rhodium-(III) is reduced to rhodium(I) and phosphorus(III) is oxidized to phosphorus(V) (reaction 142).

$$RhCl_3 + 4PPh_3 \rightarrow [Rh(PPh_3)_3Cl] + Ph_3PCl_2 \qquad 142$$

The Ph_3PCl_2 formed is unstable in 95 per cent ethanol and reacts with the water to yield triphenylphosphine oxide (reaction 143).

$$Ph_3PCl_2 + H_2O \rightarrow Ph_3PO + 2HCl \qquad 143$$

Preparation of [Rh(PPh₃)₃Br]

In order to illustrate one way in which the reactivity of the catalyst can be altered (p 73) the chloro-ligand in the Wilkinson catalyst may be replaced by a bromo-ligand. It is suggested that some pupils may prepare the chloro-catalyst and others the bromo-catalyst.

Requirements

The requirements are the same as for [Rh(PPh₃)₃Cl] together with anhydrous lithium bromide.

Method

Since the method is basically similar to that used for the chloro-catalyst the preparation of the bromo-catalyst is described more concisely.

Add a solution of triphenylphosphine (1.2 g) in hot 95 per cent ethanol (35 cm³) to a refluxing solution of rhodium trichloride (0.20 g) in hot 95 per cent ethanol (7 cm³) under nitrogen. After vigorously refluxing for a minute an orange precipitate begins to form and after another minute no further lightening in colour will occur. Add a solution of anhydrous lithium bromide (0.80 g) in hot 95 per cent ethanol (5 cm³) and reflux the solution vigorously for one hour. The orange prisms of [Rh(PPh₃)₃Br] can be collected by filtration, washed with diethyl ether (three 5 cm³ portions) and dried in a vacuum desiccator. We obtained 0.50 g, a yield of 67.5 per cent.

The catalytic hydrogenation of cyclohexene

Requirements

Cyclohexene
95 per cent ethanol
Toluene (benzene could also be used, but we specifically avoided it
  because of its toxicity)
0.1 g [Rh(PPh$_3$)$_3$Cl] or [Rh(PPh$_3$)$_3$Br]
Hydrogen
100 cm^3 gas syringe
2 three-way taps

Method

Set up the apparatus as shown using either a 100 cm^3 3-necked round-bottomed flask (*Fig. 10a*) or a 50 cm^3 2-necked pear-shaped flask (*Fig. 10b*). Pour a mixture of 95 per cent ethanol (10 cm^3) and toluene (10 cm^3) into the flask and flush with nitrogen for 15 minutes. Turn up the hydrogen flow-rate and remove joint E. Add dry [Rh(PPh$_3$)$_3$Cl] (0.1 g) (or [Rh(PPh$_3$)$_3$Br]) to the flask against the hydrogen flow by pouring it down a small filter funnel, and replace joint E as rapidly as possible. Continue vigorous stirring and hydrogen bubbling whilst the catalyst dissolves to give a very pale yellow solution — this takes about 30 minutes for [Rh(PPh$_3$)$_3$Cl] and 10 minutes for [Rh(PPh$_3$)$_3$Br]. Whilst this is taking place flush the gas syringe, and then fill it with hydrogen, using the three-way taps A and C and exhausts B and D, ensuring that no air remains in the gas syringe and associated rubber tubing nor is any inadvertently intro-

FIG. 10a. Apparatus for the hydrogenation of cyclohexene based on a 3-necked flask. **1.** B14 rubber suba-seal (see 'Cost and Availability of Materials'). **2.** B14/B24 adaptor. **3.** 100 cm^3 round-bottomed flask with 2 × B14 and 1 × B24 necks. **4.** Teflon coated magnetic follower. **5.** Water bath (*see* note 4). **6.** Magnetic stirrer. **7.** Hydrogen inlet tube dipping below the liquid surface (but not interfering with the magnetic follower). **8.** 100 cm^3 gas syringe.

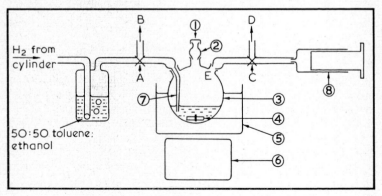

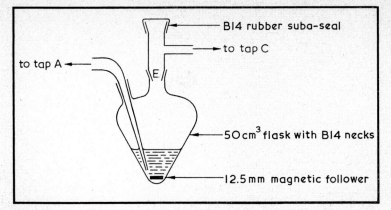

FIG. 10b. Hydrogenation apparatus based on a 2-necked pear-shaped flask. The remainder of the apparatus is as in *Fig. 10a*.

duced into the reaction flask — we found by experience that the catalyst solution is very rapidly deactivated by traces of oxygen! The accidental introduction of air can be readily recognized by a darkening of the solution.   On completion of this stage close tap C to the exhaust but leave it open between the gas syringe and the reaction vessel.   Turn tap A to seal off the flask, temporarily venting the hydrogen through B.   Shut down the hydrogen cylinder.   Fill a small graduated glass syringe (1 or 2 cm³) with 0.4 cm³ of cyclo-hexene (*see* note 5), taking care to eliminate all the air.   Inject the cyclohexene into the flask through the serum cap, whereupon the solution immediately darkens to an orange-yellow.   Continue vigorous stirring and prepare a plot of syringe reading against time (*Fig. 11*).

When all the alkene has been hydrogenated the solution will have returned to its original pale-yellow colour, and is then ready for use with further alkene.   If too much alkene has been added for the hydrogen in the 100 cm³ gas syringe the catalyst can still be regener-ated by bubbling in hydrogen until the pale-yellow colour is restored. A given batch of catalyst can be used for several runs; when we did this a precipitate finally formed in the catalyst solution, at which point a fresh solution of catalyst should be taken; furthermore until this precipitate formed we observed a gradual increase in the hydro-genation rate with successive runs.

Notes

  1. This experiment requires a good fume cupboard to remove the excess hydrogen.

  2. Because of the extreme sensitivity of the catalyst to air we found

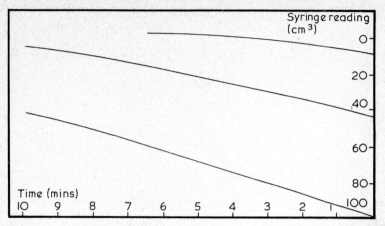

Fig. 11.   The uptake of hydrogen by cyclohexene at 19 °C in the presence of [Rh(PPh₃)₃Br] recorded on the gas syringe chart recorder.

that injection through a serum cap was the best method for introducing the alkene without any risk of air getting in.

3. The 100 cm³ gas syringe used here is described in detail by M. Rogers in *Gas syringe experiments* (Heinemann, London, 1970). If available, the chart recorder described by Rogers can also be used although the chart speed is rather high, necessitating resetting the chart every 10 minutes.   A typical chart is shown in *Fig. 11*.

4. A water-bath surrounded the flask maintaining a constant temperature by removing the heat generated by the exothermic reaction and at the same time insulating the reaction flask from the magnetic stirrer which tended to warm up during operation.   Since a non-metallic water-bath must be used (to enable the magnetic stirrer to function) we used the bottom half of a desiccator with a siphon to maintain a constant head of water (*Fig. 12*).

5. Greater accuracy in determining the amount of alkene added could be achieved if the syringe and contents were weighed (*see* Rogers, Chapter 5).   However, if this is done great care must be taken to ensure that no air is accidentally injected into the flask.

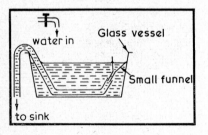

Fig. 12.   Constant head water-bath used in the hydrogenation experiments.

SCHOOL EXPERIMENTS IN ORGANOMETALLIC CHEMISTRY 81

Table 5. Typical results showing the volume of hydrogen absorbed by a given amount of alkene at 19 °C. In all cases 0.1 g of catalyst were used (i.e. 1.08×10^{-4} moles of [Rh(PPh$_3$)$_3$Cl] and 1.03×10^{-4} moles of [Rh(PPh$_3$)$_3$Br]).

Catalyst	Olefin	Vol. of olefin (cm³)	Moles of olefin	Vol. of H₂ absorbed (cm³)	Moles of H₂ absorbed	Theoretical uptake of H₂ (cm³)
[Rh(PPh$_3$)$_3$Cl]	Cyclohexene	0.40	4.0×10^{-3}	97.5	4.05×10^{-3}	96.5
	1-octene	0.63	4.0×10^{-3}	95.5	3.97×10^{-3}	96.5
	1-octene	0.63	4.0×10^{-3}	95.0	3.95×10^{-3}	96.5
[Rh(PPh$_3$)$_3$Br]	Cyclohexene	0.40	4.0×10^{-3}	100.0	4.16×10^{-3}	96.5
	1-octene	0.63	4.0×10^{-3}	99.0	4.12×10^{-3}	96.5

6. The experiment can be extended to cover a range of terminal alkenes such as 1-pentene, 1-octene *etc.* We also successfully hydrogenated pure vegetable cooking oil with this apparatus, although the rate of hydrogenation was about three times slower than with cyclohexene.

7. To ensure a uniform rate of hydrogen uptake it is essential that a uniform and vigorous rate of stirring be maintained.

Results

Some typical results are shown in Table 5. It is apparent that the observed uptake is well within the experimental error of that calculated for complete hydrogenation of the alkene. This suggests that if a known weight of an unknown terminal alkene were taken (*see* note 5 for the difficulties involved in weighing the syringe) this method could be used to determine the molecular weight of the terminal alkene and hence identify it.

In each run we recorded the rate of hydrogen uptake using the same rate of stirring. Some typical plots are shown in *Fig. 13*. These illustrate the greater effectiveness of the bromo-complex as a hydrogenation catalyst.

The catalytic hydrogenation of vegetable oil

One of the more important commercial hydrogenation reactions is the catalytic hydrogenation of vegetable oil to convert it to margarine.*

FIG. 13. Plots of the volume of hydrogen absorbed against time for cyclohexene (○) and 1-octene (□) with [Rh(PPh₃)₃Cl] (×) and [Rh(PPh₃)₃Br] (·) catalysts at 19 °C.

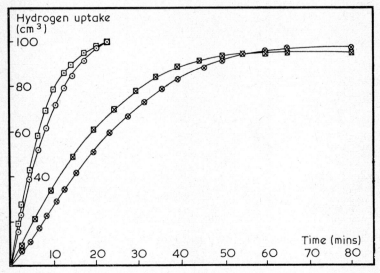

This reaction is normally carried out with a Raney nickel (heterogeneous) catalyst and since Unilever supply the catalyst and an explanatory leaflet free to schools† we compared the Wilkinson homogeneous catalyst with it, because this reaction beautifully illustrates the major drawback to homogeneous catalysts, namely the problem of separating the catalyst at the end of the reaction.

Hydrogenation of vegetable oil in the presence of [Rh(PPh₃)₃Cl]

Hydrogenation of vegetable oil in the presence of $[Rh(PPh_3)_3Cl]$

Set up the apparatus as in *Fig. 14.* Add a mixture of 95 per cent ethanol (20 cm³) and toluene (20 cm³) to the flask and stir under hydrogen for 15 minutes. Add [Rh(PPh₃)₃Cl] (0.1 g) and stir the mixture well under hydrogen until all the catalyst has dissolved to give a pale-yellow solution. Add 40 cm³ of vegetable oil (*see* note below) that has been treated with animal charcoal to remove any potential catalyst poisons and stir the mixture under hydrogen at room temperature for eight hours — *note* filtering off the animal charcoal is a slow process owing to the viscosity of the vegetable oil and should preferably be done the day before the hydrogenation. After about six hours the solution becomes cloudy due to the formation of a solid fatty product.

After about eight hours treat the mixture with water (to dissolve

FIG. 14. Apparatus for the hydrogenation of vegetable oil in the presence of [Rh(PPh₃)₃Cl].

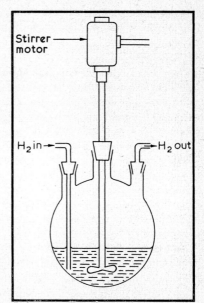

Stirrer motor

H₂ in →

← H₂ out

* Manufacturers use a wide range of vegetable oils to produce margarine hence the product of this experiment will be only one of the hydrogenated vegetable oils present in margarine.

† *The hydrogenation of soft oils,* Unilever Laboratory Experiment No. 3, available to science teachers from Unilever Educational Section, Unilever House, Blackfriars, London EC4.

some of the ethanol) and diethyl ether and separate the ether layer,
dry it with anhydrous sodium sulphate and distil off the ether on a
water-bath to leave a residue which should partly solidify on cooling
in an ice-bath.

The product would be margarine but it is contaminated with the
catalyst and toluene. Can you devise an inexpensive method to
remove them completely? If not you will appreciate the problem
that a margarine producer would be faced with.

Hydrogenation of vegetable oil in the presence of Raney nickel

Obtain the Unilever leaflet and the sample of catalyst. Hydrogenate
the vegetable oil according to the instructions and observe how easy it
is to simply filter off the catalyst at the end of the reaction.

Note

In this experiment we used cooking oil obtained from a local grocers.
Alternatively olive oil or corn oil could have been used, although
both are more expensive.

Experiment II.   The homogeneous oxidation of alkenes in the presence of palladium(II) salts (the Wacker process)

This experiment was developed in association with Dr A. W. Shaw of
The South Wilts Grammar School for Girls, Salisbury.

Introduction

In this experiment the student observes the essential features of the
Wacker Process for the homogeneous oxidation of alkenes to carbonyl
compounds described in detail in Chapter 5 (pp 50–52). Most of the
essential parts of this chemistry are carried out in a test-tube by the
student. During the experiment the student prepares some new
organic compounds and characterizes these into type (*i.e.* aldehyde or
ketone) before going on to prepare a derivative which he melts in
order to determine which aldehyde or ketone was formed. He also
demonstrates the importance of obtaining a homogeneous solution in
that no reaction occurs between aqueous Na_2PdCl_4 and higher alkenes
until the alkene is induced to dissolve in the water either by adding
alcohol or by heating the mixture.

Cost and availability of materials

The palladium(II) salt used throughout this work is sodium tetra-
chloropalladate(II), Na_2PdCl_4, which may either be purchased directly
or prepared from $PdCl_2$ and sodium chloride as in reaction 144.

$$PdCl_2 + 2NaCl \rightarrow Na_2PdCl_4 \qquad 144$$

Na$_2$PdCl$_4$* can be purchased either from Engelhard (p 79) who charge† 50p per gm (exclusive of VAT) with a minimum order of £5 or from Johnson Matthey (p 79) who charge† £1.30 per gm, with a minimum order of £5, or £6.61 per 10 g (both exclusive of VAT). PdCl$_2$ is available as a solid either from these suppliers or more expensively, but with a minimum order of 1 g, from BDH or Hopkin and Williams at a cost† of about £1.40 per gm, or as a 2 per cent aqueous solution (from Hopkin and Williams) which could then be treated direct with the equivalent amount of sodium chloride to give a solution of Na$_2$PdCl$_4$.

Liquid alkenes can be purchased from BDH, Koch-Light or Ralph N. Emanuel. Gaseous ethene can be purchased from BDH or British Oxygen (Special Gases Division, Deer Park Road, London SW19). Because of the complexity of prices for hiring the lecture bottles we have not given a price for ethene here. Furthermore we assume that most schools will prefer to prepare it as required, and accordingly give a preparation which we have found to be very satisfactory.

The preparation of ethene

Ethene may be conveniently prepared in the laboratory by the dehydration of ethanol. Older textbooks tend to recommend the use of concentrated sulphuric acid or phosphoric acid, but such preparations tend to be complex and take a considerable time to set up. Accordingly we used the catalytic dehydration of ethanol over hot aluminium oxide since, using the apparatus shown in *Figs 15* and *16*, about 300 cm^3 of ethene could be prepared and collected in 20 minutes.

FIG. 15. Apparatus for the preparation of ethene.

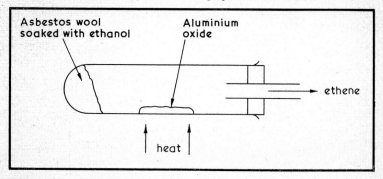

* Commercially Na$_2$PdCl$_4$ is referred to as 'sodium tetrachloropalladite (Na$_2$PdCl$_4$)'.
† Note all costs are approximate (for early 1974) and are given purely to give an idea of the overall costs. They are likely to change at any time.

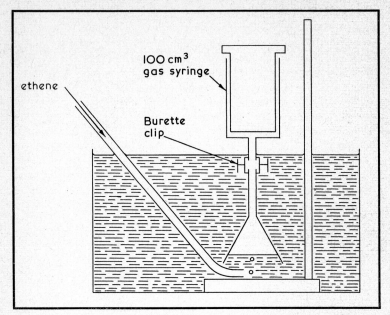

FIG. 16.   Apparatus for the collection of ethene.

Place a pad of asbestos wool soaked in ethanol at the bottom of a boiling tube clamped in a horizontal position as shown in *Fig. 15*. Place a heap of aluminium oxide (technical grade) at the top of the tube and connect the tube to the collection apparatus shown in *Fig. 16*. Heat the aluminium oxide and allow the stream of ethene to collect in the funnel which should be clamped (to avoid it floating away!). Any ethanol that is carried over dissolves in the water. Fill a 100 cm³ gas syringe with ethene and when it is full reseal the funnel.

The oxidation of alkenes

(i) Oxidation of ethene to acetaldehyde by Na₂PdCl₄ in a test-tube

Reaction 81 (p 50) may be illustrated as follows: Prepare a solution of Na_2PdCl_4 (0.1 g) in water (75 cm³). Slowly bubble about 30 cm³ of ethene from a gas syringe into a test-tube containing 5 cm³ of the palladium(II) solution and observe the change in colour from light-brown to black due to the formation of finely divided palladium metal. Warm the test-tube in a hot water-bath and observe the coagulation of the palladium metal into lumps. Smell the test-tube and observe the apple-like smell of acetaldehyde — if acetaldehyde is available, then use it for comparison. If a stronger solution of the palladium(II) salt is used (*e.g.* 0.1 g of Na_2PdCl_4 in 5 cm³ of water) then a palladium mirror is formed on the walls of the tube, rather than finely divided palladium.

(ii) Oxidation of ethene to acetaldehyde by Na₂PdCl₄ in the presence of copper(II) chloride and air

Reactions 81 and 86–88 (pp 51–52) may be illustrated using the apparatus shown in *Fig. 17.* Pour a solution of Na_2PdCl_4 (0.1 g) and copper(II) chloride (2.0 g) in water (50 cm³) into the 3-necked flask. Bring the water bath to the boil and slowly bubble in 100 cm³ of ethene from a gas syringe. After heating for a further 10 minutes connect up the water pump and draw air through the apparatus for 10 minutes to reoxidize the copper(I) chloride and also to carry the acetaldehyde over into the boiling tube containing Tollens' reagent (*see* note 1) which is surrounded by an ice-bath. Disconnect the water pump, remove the boiling tube containing Tollens' reagent and warm it in the water-bath to develop a silver mirror, confirming the presence of an aldehyde.

The experiment can be repeated many times using the same catalyst solution by the alternate passage of ethene and air. If instead of carrying out a single aldehyde test it is desired to let each student carry out the test and also smell the acetaldehyde, the boiling tube should be half-filled with water only. The resulting solution of acetaldehyde obtained after several cycles of ethene and air can then be divided amongst the students.

FIG. 17.   Apparatus for the preparation of acetaldehyde.

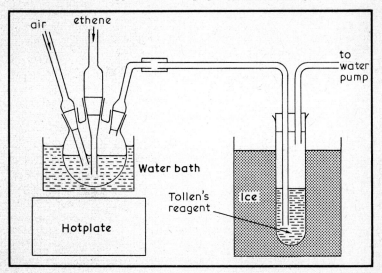

Notes

1. Tollens' reagent *must not be kept*, as it slowly forms very explosive substances. It must be freshly prepared, and washed away after use. Tollens' reagent is prepared by treating 2 cm³ of a 5 per cent aqueous silver nitrate solution with one drop of dilute sodium hydroxide solution and adding dilute (*ca* 2 per cent) aqueous ammonia dropwise until the precipitate just redissolves; excess of ammonia must be avoided. Aldehydes give a silver mirror, ketones (except α-diketones and α-hydroxyketones) do not.

2. The palladium : copper ratio is very important. Too much copper stops the reaction whilst too little leads to the precipitation of palladium metal.

3. The yield of acetaldehyde in this experiment, whilst adequate for demonstration purposes is low relative to the industrial process. This is due to the inefficient absorption of the ethene by the catalyst solution in the system compared to the gas absorption towers used commercially.

(iii) Oxidation of liquid alkenes by Na_2PdCl_4 in a test-tube

Higher alkenes such as hex-1-ene and oct-1-ene are oxidized in a similar way to ethene to yield methyl ketones (reaction 145, R = alkyl).

$$RCH{=}CH_2 + PdCl_2 + H_2O \rightarrow R\underset{\underset{O}{\|}}{C}CH_3 + Pd^0 + 2HCl \qquad 145$$

However, to get the reaction to go, the alkene must dissolve in the water to some extent. This can either be achieved by heating or by the addition of a solvent such as methanol in which both the alkene and water dissolve. Ethanol is not used in the following experiments because, although it is in other ways ideal, it is itself oxidized by palladium(II) to acetaldehyde and in any qualitative test the purist would be right to object that our proof of obtaining a carbonyl product was spurious. Methanol is also oxidized by palladium(II) but only very slowly.

Prepare a solution of Na_2PdCl_4 (0.1 g) in water (75 cm³). Give each student two test-tubes containing 2 cm³ of this solution. To each add 1 cm³ of a particular alkene and shake. No reaction will be observed if that alkene is hex-1-ene, oct-1-ene or cyclohexene, but with styrene a faint darkening may be visible at the interface of the two layers. To one of the test-tubes add methanol (2 cm³) and observe the immediate darkening of the solution and precipitation of palladium metal — carry out at least one blank test of adding methanol to the Na_2PdCl_4 solution in the absence of alkene to confirm that the methanol does not readily reduce Na_2PdCl_4 to palladium metal under the present conditions.   Heat the other test-tube in a boiling water-bath to promote the reaction and again observe the deposition of palladium metal over a period of 20 minutes.   As shown below in part *iv*, hex-1-ene is oxidized to hexan-2-one (*n*-butylmethylketone), oct-1-ene to octan-2-one (*n*-hexylmethylketone) and styrene to a mixture of acetophenone ($C_6H_5COCH_3$) and phenylacetaldehyde ($C_6H_5CH_2CHO$).   Cyclohexene is oxidized to cyclohexanone.

(iv) Oxidation of liquid alkenes by Na_2PdCl_4 and copper(II) chloride and identification of the products

In this experiment liquid alkenes are oxidized by a solution of Na_2PdCl_4 and copper(II) chloride and the product of the oxidation identified, first as an aldehyde or ketone by suitable tests and, secondly, as a specific compound by preparing derivatives and recording their melting-points.   Although we recognize that few schools will have it, we recommend the use of oct-1-ene in preference to the commoner hex-1-ene because (*a*) the reaction is faster, (*b*) the lower volatility of oct-1-ene is advantageous, (*c*) the 2,4-dinitro-phenylhydrazone derivative melts below 100 °C and so its melting-point can be determined in a water-bath and (*d*) ethanol rather than the less common methanol can be used as the solvent.   We do not recommend attempting to reoxidize the copper(I) chloride with air, as we had little success with it.

Oct-1-ene.   Gently reflux a solution of Na_2PdCl_4 (0.1 g), copper(II) chloride (6 g), oct-1-ene (10 cm³), ethanol (40 cm³) and water (35 cm³) for half-an-hour.   The initial greenish-yellow colour deepens to a brownish-yellow and later a white precipitate of copper(I) chloride can be seen.   Finally a palladium mirror forms. Distil the mixture and test the first fraction which contains ethanol, water and the ketone for a ketone as follows:

(*i*) Smell it and observe a smell reminiscent of an ester such as amyl acetate.

(*ii*) Treat a portion with a solution of 3,5-dinitrobenzoic acid (0.1 g) (*see* note 1) in 2 M sodium hydroxide solution (1 cm³) to produce a purple colouration — this test is given by ketones but not aldehydes; if a sample of octan-2-one (*n*-hexylmethylketone) is available and is to be used for comparison, 1 cm³ of ethanol should be added to ensure solubility (the sample already contains ethanol).

(*iii*) Treat a portion with Tollens' reagent (*see* above, part *ii*, note 1) and note the failure to produce a silver mirror indicating the absence of an aldehyde.

(*iv*) Prepare the 2,4-dinitrophenylhydrazone derivative (*see* note 2) and determine its melting-point (58 °C) confirming that the ketone is octan-2-one (*n*-hexylmethylketone).

Notes

1. 3,5-Dinitrobenzoic acid can either be purchased commercially (it costs* approximately £1 per 100 g) or it may be prepared by nitration of benzoic acid† in which case it provides a useful alternative to the nitration of benzene often carried out during A-level studies.

2. The 2,4-dinitrophenylhydrazone derivative is prepared as follows: suspend 1 g of powdered 2,4-dinitrophenylhydrazine in methanol (30 cm³). Stir and cautiously add concentrated sulphuric acid (2 cm³). If necessary, filter the solution whilst it is still warm. Add 1 cm³ of this solution to a portion of the distillate. On shaking for a few minutes and scratching, if necessary, the yellowish-orange derivative precipitates. Filter off the product at the water-pump, wash with a few drops of methanol and allow the product to dry in air before determining its melting-point.

3. If desired the copper(I) chloride may be filtered off, washed successively with sulphurous acid (to prevent air oxidation of Cu(I) to Cu(II) occurring), ethanol and diethyl ether, dried and weighed. Typically 6 g of $CuCl_2.2H_2O$ used straight from the bottle yielded 3.3 g of CuCl as compared to a theoretical yield of 3.5 g expected if the copper(II) chloride had been dry and exactly as represented by the formula, and if all the copper(II) had been reduced to copper(I).

Hex-1-ene. Treat hex-1-ene in exactly the same way as oct-1-ene except that methanol should be used in place of ethanol and refluxing should continue for one hour. Although the reason why methanol is necessary is not fully clear it may be connected with the fact that methanol boils at a lower temperature than hex-1-ene, so that in methanolic solution more of the hex-1-ene remains in solution rather than being volatilized in the gas phase. The presence of a ketone was shown in exactly the same way as with oct-1-ene. The 2,4-dinitrophenylhydrazone derivative of hexan-2-one (*n*-butylmethyl-ketone) melts at 99 °C.

Styrene. Treat styrene in exactly the same way as oct-1-ene using ethanol as the alcohol. Considerable polymerization to polystyrene will be observed, a reaction known to be catalysed by palladium salts.

---

* Note all costs are approximate (for early 1974) and are given purely to give an idea of overall costs. They are likely to change at any time.
† For preparative details see either F. G. Mann and B. C. Saunders, *Practical organic chemistry*, Fourth Edition, Longmans, London, 1960, p 240 or A. I. Vogel, *A textbook of practical organic chemistry*, Third Edition, Longmans, London, 1956, p 770.

Distillation yields a distillate that, (*i*) gives a silver mirror with Tollens' reagent indicating the presence of an aldehyde; and (*ii*) gives a faint purple colouration with 3,5-dinitrobenzoic acid indicating the presence of a ketone. This is consistent with the presence of both phenyl-acetaldehyde and acetophenone. In agreement with this our 2,4-dinitrophenylhydrazone derivative melted at 233 °C, which is between that of the derivatives of phenylacetaldehyde (121 °C) and aceto-phenone (250 °C).

Cyclohexene. We were unable to find suitable conditions for the oxidation of cyclohexene to cyclohexanone in the presence of copper(II) chloride.

Arrangement of the experimental work

Whilst recognizing that it is presumptuous to suggest how a class should be organized we feel that some suggestions may be in order. We suggest that each student prepares a sample of ethene which he keeps trapped in his funnel until a gas syringe is available. Each student uses this to oxidize it to acetaldehyde as in part *i*. Mean-while the teacher sets up part *ii* and either carries out part *ii* as a class demonstration or carries part *ii* on sufficiently long to give each student a sample on which to carry out his own Tollens' test. We then suggest that each student carries out the test-tube oxidation of liquid alkenes (part *iii*) whilst the oxidation of liquid alkenes in the presence of Na_2PdCl_4 and $CuCl_2$ and identification of the products is carried out by the teacher as a demonstration experiment. With careful planning all the work can be fitted into a 2-hour period except for the melting-point of the derivative, which must be allowed to dry first. For a class of 15 this scheme would involve the use of 0.4 g of Na_2PdCl_4.

Experiment III.   The preparation of ferrocene

This experiment was developed in association with Mrs G. Temple-Nidd of The Southampton College for Girls.

Introduction

Ferrocene has always been a key compound in organometallic chemistry because recognition of its unique structure led to a tre-mendous upsurge of interest in the subject. It is additionally one of the most beautifully crystalline compounds in chemistry and the long crystals formed during its sublimation are calculated to excite even the least interested student. In the course of the experiment the aromatic properties of ferrocene are exemplified by its Friedel–Crafts acetylation. This leads to a mixture of ferrocene and acetylferrocene which can be separated by thin layer chromatography. Besides demonstrating a simple application of this valuable technique, including the use of a stain, this experiment also demonstrates an important aspect of organometallic chemistry, namely that although it may well be possible to write an equation for a reaction this is no guarantee that that reaction can be made to go to completion.

Requirements
Dicyclopentadiene (approximate cost 88p per 1 kg)
Dimethylsulphoxide (approximate cost £1.15 per 500 cm³)
Iron(II) chloride, $FeCl_2.4H_2O$ (approximate cost 50p per 250 g)
Potassium hydroxide flake (technical)
Acetic anhydride
Orthophosphoric acid
(These chemicals can be obtained from such suppliers as BDH, Hopkin & Williams and Koch-Light.)

Experimental
Preparation of cyclopentadiene
Cyclopentadiene readily dimerizes at room temperature to dicyclo-pentadiene by a Diels–Alder reaction (reaction 146)

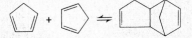

146

FIG. 18a. Traditional apparatus for 'cracking' dicyclopentadiene in which the vertical condenser is maintained at 56 °C with refluxing acetone.

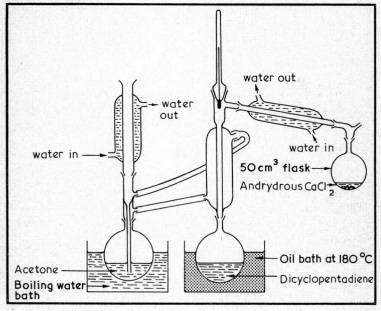

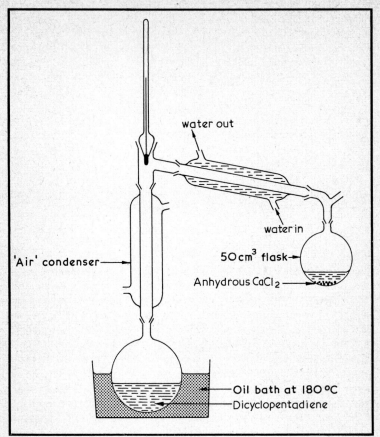

FIG. 18b. Simplified apparatus for the 'cracking' of dicyclopentadiene in which the vertical condenser is left open to the air.

and at higher temperatures further polymerization occurs.   Accordingly cyclopentadiene is sold commercially as its dimer and this must be 'cracked'.   The traditional apparatus for 'cracking' dicyclopentadiene, shown in *Fig. 18a*, requires a considerable amount of standard glassware together with one non-standard piece.   Accordingly in this work we have simplified the apparatus (*Fig. 18b*) to bring it within the reach of a school budget yet still give a satisfactory preparation of cyclopentadiene.

On arrival in the morning switch on the hotplate.   Set up the apparatus as shown in *Fig. 18b*.   Put approximately 60 cm³ of dicyclopentadiene in the flask and allow the oil-bath to heat up to 180 °C.   At first there is considerable frothing, but this subsides after a time and the product distills over steadily.   Collect the product, which slowly distills off at 42–44 °C, in a flask containing anhydrous calcium chloride. This is a fairly lengthy process and should be done by the teacher; about 10 cm^3 will distill off in the first $2\frac{1}{2}$ hours and a further 20 cm^3 in the next 3 hours. If possible the 'cracking' should be done on the day of the experiment as the dicyclopentadiene dimerizes at a significant rate at room temperature. If the cyclopentadiene is required for a morning practical session then it must be stored overnight in the freezing compartment of a refrigerator, but this is not recommended.

Preparation of ferrocene

Ferrocene is prepared by treating iron(II) chloride with potassium cyclopentadienide, made by the reaction of cyclopentadiene with potassium hydroxide (reactions 147 and 148).

$$C_5H_6 + KOH \rightarrow K^+C_5H_5^- + H_2O \qquad\qquad 147$$

$$2K^+C_5H_5^- + FeCl_2 \rightarrow (C_5H_5)_2Fe + 2KCl \qquad\qquad 148$$

The water formed in reaction 147 is removed by having an excess of potassium hydroxide present. Normally the reaction is carried out in 1,2-dimethoxyethane (MeOCH$_2$CH$_2$OMe) under nitrogen because potassium cyclopentadienide is air sensitive. However, in this preparation nitrogen is not used (because of its expense to schools) and the bulk of the air is kept out by using diethylether which is rather volatile and blankets the reaction. Since ferrocene is endothermic with a heat of formation of $+141$ kJ mol^{-1}, the driving force for the reaction is provided by the formation of potassium chloride (heat of formation -436 kJ mol^{-1}).

Assemble the apparatus shown in *Fig. 19* in a fume cupboard. Weigh out 6.5 g of iron(II) chloride (FeCl$_2$.4H$_2$O) (*see* note 1), grind to a fine powder and leave it to dissolve in 25 cm^3 of dimethylsulphoxide with occasional stirring (*care* — although the toxicity of dimethylsulphoxide is not well known, it is known to be a mild swelling agent for proteins and therefore contact with the skin should be avoided*). Place flake potassium hydroxide (25 g) (*see* note 2) in the conical flask shown in *Fig. 19*, add diethylether (60 cm^3) and stir for 10 minutes to dissolve as much as possible. Add cyclopentadiene (5.5 cm^3) and stir for a further 10 minutes before adding the solution of iron(II) chloride in dimethylsulphoxide dropwise over a period of 45 minutes. During the addition an exothermic reaction occurs and the ether may boil; if it does, release the Bunsen valve occasionally. Continue stirring for a further 30 minutes.

Decant the ethereal layer and wash twice with 25 cm^3 of 2 M

---

* In case of contact wash well with water, with which dimethylsulphoxide is readily miscible.

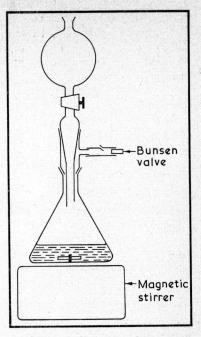

FIG. 19. Apparatus for the preparation of ferrocene.

hydrochloric acid (to neutralize the alkali) and twice with 25 cm³ of water.   Carefully evaporate the ether in a fume cupboard to deposit orange-brown crystals of ferrocene (1.6 g) in 25 per cent yield.   Put 0.5 g aside for conversion to acetylferrocene and purify the remainder by placing them in the crystallizing dish shown in *Fig. 20*.   Place a petri cover over this dish and warm it on a hotplate in a fume cupboard — objectionable fumes of ferrocene will fill the laboratory if this is done on the open bench.   Long golden crystals of pure ferrocene will be formed on the lid and may be removed from time to time.

Notes

1. The iron(II) chloride used should be from as fresh a bottle as possible, since the yield deteriorates as the iron(II) chloride ages, for

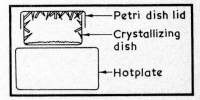

FIG. 20. Apparatus for the sublimation of ferrocene.

example with one old obviously oxidized sample a yield of 15 per cent was obtained. An attempt to replace iron(II) chloride with iron(II) sulphate, which is more readily available in schools and also less readily oxidized by air, gave an overall yield of 4 per cent so we do not recommend its use!

2. An attempt to replace potassium hydroxide by sodium hydroxide, which is more readily available in schools, gave an overall yield of 2 per cent. This disappointing result was due to the low solubility of sodium hydroxide in diethyl ether.

3. When this synthesis was carried out using double these quantities the yield improved markedly to 53 per cent (6.4 g).

Preparation of acetylferrocene

Ferrocene shows a number of aromatic properties, one of which is its ability to undergo a Friedel–Crafts reaction. Typically this is achieved using the Lewis acid aluminium trichloride as the catalyst. However, phosphoric acid is more convenient as a catalyst and is used in this preparation which is based on equation 149.

$$(CH_3CO)_2O + (C_5H_5)_2Fe \xrightarrow{H_3PO_4}$$
$$[(C_5H_5)Fe(C_5H_4COCH_3)] + CH_3COOH \quad 149$$

The product obtained is a mixture of ferrocene and acetylferrocene which is separated using thin layer chromatography.

Dissolve crude ferrocene (0.5 g) in acetic anhydride (5 cm³) in a small flask provided with a silica gel guard tube and add dropwise with shaking 1 cm³ of orthophosphoric acid. Heat in a hot water-bath for 30 minutes. Pour the mixture onto crushed ice (20 g) and when the ice has melted neutralize the solution with solid sodium hydrogen carbonate before filtering off the brownish-yellow solid at the pump. Dry overnight in the open.

Thin layer chromatography

Acetylferrocene can be separated from ferrocene using thin layer chromatography. For those experienced in this technique this will present no problems. For those who have not previously used it we recommend using one of the standard kits such as an Eastman Chromagram kit (model 104) which is available from Eastman–Kodak (Kirkby, Liverpool) at an approximate cost of £4.50 (excluding VAT) with further replacement thin layer sheets costing approximately £1.27 (excluding VAT) per box of 25.

Prepare 1 per cent solutions of acetylferrocene and ferrocene (for comparison) in diethylether. Spot the plates and run the chromatogram for approximately 20 minutes using a mixture of 10 per cent (by volume) ethyl acetate and 90 per cent petroleum ether (bp 60–80 °C) as

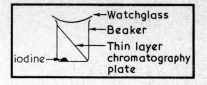

FIG. 21. Apparatus for exposing thin layer chromatography sheets to iodine vapour.

eluant.   Dry the sheet and develop the spots by exposing the sheet to iodine vapour for about 20 minutes using the apparatus shown in *Fig. 21*.   Mark the spots, because they fade after a while, and determine the R_f values of ferrocene and acetylferrocene.

$$R_f \text{ value} = \frac{\text{(distance of the centre of the spot from the origin)}}{\text{(distance of the eluant front from the origin)}}$$

Experiment IV.   The preparation of tetraphenyllead

This experiment was developed in association with Mrs G. Temple-Nidd of The Southampton College for Girls and Mr T. G. B. Hackston of Barton Peveril College, Eastleigh.

Introduction

In choosing an example of an alkyl or aryl compound we were guided by two principles, firstly a desire not to neglect non-transition metal chemistry and secondly a desire to choose a compound of some applicability.   The most obvious example was clearly tetraethyllead, the well known anti-knock additive in petrol.   However, this compound is firstly far too toxic to handle in a school and secondly it is a liquid and difficult to characterize.   Accordingly we have developed a synthesis of tetraphenyllead, which is similar to tetraethyllead, but far less toxic.   It is a crystalline compound which facilitates its characterization.

The starting material is lead(II) chloride and during the course of the reaction this disproportionates to lead(IV) and metallic lead, which is precipitated (reaction 150).

$$2PbCl_2 + 4PhMgBr \rightarrow PbPh_4 + Pb^0 + 4MgBrCl \qquad 150$$

The white crystalline product is characterized by its melting-point and by decomposition in concentrated sulphuric acid followed by standard qualitative tests to show the presence of lead.

Requirements

Magnesium turnings
Diethyl ether (dried over anhydrous calcium chloride)
Bromobenzene
Toluene (dried over anhydrous calcium chloride)

Lead(II) chloride
Ammonium chloride
Chloroform

Experimental
The preparation of phenylmagnesium bromide

All the apparatus should be dried in an oven before use. Place magnesium turnings (3 g), bromobenzene (4 cm³) and diethyl ether (8 cm³) in a 250 cm³ round-bottomed flask containing a magnetic follower and fitted with a reflux condenser. Add a crystal of iodine; this provides a nucleus on the magnesium surface for the reaction to start from, so it should be kept in one place and not allowed to move about unnecessarily. Warm the flask gently with the hand until the reaction starts, as indicated by bubbling. Fit a dropping funnel and Bunsen valve to the top of the condenser (or to a second inlet if a multi-necked flask is available) and from this add, over a period of 20 minutes, a solution of bromobenzene (10 cm³) in diethyl ether (50 cm³) at such a rate that the ether refluxes gently. Reflux for a further 30 minutes over a warm water-bath to complete the reaction. At this stage the solution is cloudy and brown and little, if any, magnesium should remain.

The preparation of tetraphenyllead

Lead(II) chloride should be stored overnight in a desiccator containing silica gel to remove any surplus moisture. To the phenylmagnesium bromide solution prepared above add 40 cm³ of dry toluene followed by 14.5 g of lead(II) chloride. Place the flask in an oil-bath over a magnetic stirrer hotplate and reflux with stirring, either for 5 hours, or, if less time is available, for as long as possible. If refluxing is for significantly less than 5 hours allow the mixture to stand at room temperature overnight.

Hydrolyse the mixture by pouring it into a beaker containing a mixture of ammonium chloride (27 g), crushed ice (125 g) and water (125 cm³). Stir for 15 minutes. Decant off the ethereal layer and filter off the greyish solid from the aqueous layer using a Buchner funnel. Extract the solid twice with 50 cm³ portions of hot chloroform and dry the combined extracts over anhydrous magnesium sulphate for a few minutes. Filter, and evaporate off the chloroform to yield a virtually white product which may be recrystallized from carbon tetrachloride or toluene to give 2 g (13.4 per cent yield) of fluffy white needles of tetraphenyllead.

The characterization of tetraphenyllead

Record the melting-point of the recrystallized material. It should be 224 °C. Show that the product contains lead as follows:

Warm 0.1 g of the product with concentrated sulphuric acid (5 cm^3) in a 100 cm^3 beaker until only a slight precipitate of lead sulphate appears. Carefully pour the contents of the beaker into water (25 cm^3) when a considerable precipitate of lead sulphate is formed. Allow the precipitate to settle, decant off the supernatant liquid and dissolve the precipitate by adding a large excess (*ca* 3 g) of ammonium acetate (or sodium acetate) and some water. Test the solution for the presence of Pb^{2+} as follows:

(*a*) Add a solution of potassium chromate to obtain a copious yellow precipitate of lead(II) chromate.

(*b*) Add a solution of potassium iodide to obtain a yellow precipitate of lead(II) iodide that is soluble in excess potassium iodide but precipitated on further dilution.

(*c*) Treat the solution with hydrogen sulphide to obtain a black precipitate of lead(II) sulphide.

Experiment V.   The preparation of triphenylphosphine and some of its complexes with transition metals

This experiment was developed in association with Mr T. G. B. Hackston of Barton Peveril College, Eastleigh.

Introduction

Although not strictly an organometallic compound itself, triphenylphosphine forms complexes with most transition metals and these complexes are widely used as starting materials for the preparation of transition metal organometallic complexes. Furthermore the two routes for the preparation of triphenylphosphine (reactions 151 and 152)

$$3PhMgBr + PCl_3 \rightarrow PPh_3 + 3MgBrCl \qquad\qquad 151$$

$$3PhLi + PCl_3 \rightarrow PPh_3 + 3LiCl \qquad\qquad 152$$

are typical of the routes used to prepare metal–aryl complexes (pp 16–17). In this experiment triphenylphosphine is prepared using phenylmagnesium bromide because the greater sensitivity of phenyllithium to air necessitates the use of nitrogen, or argon, in order to obtain reasonable yields from reaction 152. The triphenylphosphine is used here to prepare complexes of cobalt(II), nickel(II) and copper(I) and in Experiment I (pp 75–77) to prepare two rhodium(I) complexes. Two of these complexes, [Co(PPh$_3$)$_2$Cl$_2$] and [Ni(PPh$_3$)$_2$Cl$_2$] are treated with phenylmagnesium bromide to yield unstable aryl complexes [Co(PPh$_3$)$_2$Ph$_2$] and [Ni(PPh$_3$)$_2$Ph$_2$].

Requirements

Magnesium turnings
Diethyl ether (dried over anhydrous calcium chloride)
Toluene (dried over anhydrous calcium chloride)
Bromobenzene
Phosphorus trichloride
Cobalt(II) chloride ($CoCl_2.6H_2O$)
Nickel(II) chloride ($NiCl_2.6H_2O$)
Copper(II) chloride ($CuCl_2.2H_2O$)

Experimental

The preparation of triphenylphosphine

This preparation should be carried out in a fume-cupboard.

Prepare a solution of phenylmagnesium bromide in ether as described on p 98 using 5.1 g of magnesium, 22 cm³ of bromobenzene and 100 cm³ of dry diethyl ether. Stir the phenly-magnesium bromide solution on a magnetic stirrer whilst cooling it with an ice-salt bath. Add a solution of phosphorus trichloride (3.5 cm³) in dry diethyl ether (30 cm³) dropwise over a period of 30 minutes. A vigorous reaction occurs as each drop enters and a yellow solid may begin to separate, if it does so then the addition is being made too rapidly.

Prepare a solution of hydrochloric acid (11 cm³ of concentrated acid in 80 cm³ of water) and place this in a 1000 cm³ beaker surrounded by an ice-salt freezing mixture. Slowly pour the contents of the reaction flask into the acid, a little at a time, with vigorous stirring. When the hydrolysis is complete the bulk of the triphenylphosphine will be in the ether layer. Decant off the ether layer, extract the aqueous layer with a further 25 cm³ of ether and dry the combined ether fractions over anhydrous sodium sulphate until the ether is clear. Filter off and evaporate the filtrate to obtain a yellow semiliquid residue which solidifies overnight to yield 7 g of crude triphenylphosphine. Recrystallize the product from ethanol, retaining a few particles of the solid to seed the cooled ethanolic solution. Filter off the white crystalline triphenylphosphine in a Buchner funnel, washing the product with 1 cm³ portions of ethanol until all traces of yellow colour have disappeared. Our yield was 4 g (38 per cent). Melting-point 79–80 °C.

Preparation of triphenylphosphine complexes

(*a*) [$Co(PPh_3)_2Cl_2$]. Dissolve finely powdered pink cobalt(II) chloride ($CoCl_2.6H_2O$) (0.60 g; 2.5 mmole) in ethanol (7 cm³) to obtain a rich blue solution. Add a solution of triphenylphosphine (1.31 g; 5 mmole) in ethanol (6 cm³) to obtain an immediate precipitate of bright blue crystals which are collected in a Buchner funnel and washed with a little ethanol before drying.   Yield 1.15 g; melting-point 235–240 °C.

(b) $[Ni(PPh_3)_2Cl_2]$.   Stir nickel(II) chloride ($NiCl_2.6H_2O$) (0.60 g; 2.5 mmole) in water (0.5 cm³) and glacial acetic acid (12.5 cm³) to obtain a pale-green suspension.   Add, with stirring, a solution of triphenylphosphine (1.31 g; 5 mmole) in glacial acetic acid (6 cm³) to obtain a very dark-green crystalline precipitate which becomes almost black on standing.   Filter off in a Buchner funnel and wash with a little acetic acid before drying.   Yield 1.31 g; melting-point 260 °C (with decomposition).

(c) $[Cu(PPh_3)Cl]_4$.   Dissolve copper(II) chloride ($CuCl_2.2H_2O$) (0.57 g; 3.4 mmole) in ethanol (18 cm³) to obtain a green solution. Add a solution of triphenylphosphine (1.31 g; 5 mmole) in ethanol (10 cm³) and heat the mixture under reflux for a short time until the green colour disappears indicating that the copper(II) has been reduced to copper(I) by the triphenylphosphine (cf. reactions 142 and 143, p 77).   At the same time an off-white solid is deposited, which is filtered off in a Buchner funnel and washed with ethanol (5 cm³) to obtain fine pale-beige crystals.   Yield 1.34 g; melting-point 238–240 °C.

The presence of copper in the complex can be confirmed as follows: Heat 0.1 g of the complex with 1 cm³ of concentrated perchloric acid* and 1 cm³ of concentrated nitric acid in a small beaker until all the solid has dissolved.   This yields a blue-green solution typical of copper(II).   Add 5 cm³ of water and filter.   Divide the filtrate into two parts and treat as follows:

1. To one part add 3 cm³ of concentrated aqueous ammonia to obtain the deep blue colour of $[Cu(NH_3)_4]^{2+}$.

2. Pass hydrogen sulphide into the other part to obtain a brownish-black precipitate of copper(II) sulphide.

Reaction of $[Co(PPh_3)_2Cl_2]$ and $[Ni(PPh_3)_2Cl_2]$ with PhMgBr

$[Co(PPh_3)_2Cl_2]$ and $[Ni(PPh_3)_2Cl_2]$ react with phenylmagnesium-bromide to give unstable phenyl complexes that decompose slowly in the absence of air and rapidly in the presence of air (reaction 153).

$$[M(PPh_3)_2Cl_2] + 2PhMgBr \xrightarrow{\text{(M=Co, Ni)}} [M(PPh_3)_2Ph_2] + 2MgBrCl \quad 153$$

In this experiment the formation of the phenyl complex is observed by the discharge of the colour of the chloro-complex on adding the

---

* Concentrated sulphuric acid can be used instead of concentrated perchloric acid, but it is less efficient.

phenylmagnesium bromide solution.   On exposure to air decomposition with the formation of green solids is observed in both cases.
   Prepare a solution of phenylmagnesium bromide in ether as
described on p 98 using 1.15 g of magnesium, 5.0 cm^3 of bromobenzene and 75 cm^3 of dry diethyl ether.

Reaction between [Co(PPh₃)₂Cl₂] *and PhMgBr.*   Prepare a fine blue
suspension of [Co(PPh$_3$)$_2$Cl$_2$] (1.0 g; 1.5 mmole) in dry toluene
(20 cm^3) in a conical flask provided with a magnetic follower and an
air condenser.   Pipette out 5 cm^3 of the phenylmagnesium bromide
solution (3.0 mmole) and add it to the stirred cobalt(II) solution.
The colour immediately darkens to deep brown.   Divide the solution
into two parts.   Expose one part to air and observe the deposition of
a green solid.   Treat the other part with animal charcoal (to remove
any impurities) and filter.   The first few drops of the filtrate are
almost colourless, as expected for [Co(PPh$_3$)$_2$Ph$_2$], but later decomposition to yield a green product occurs.

Reaction between [Ni(PPh₃)₂Cl₂] *and PhMgBr.*   Carry out exactly the
same operations as described for cobalt using 1.0 g (1.5 mmole) of
[Ni(PPh$_3$)$_2$Cl$_2$] and 5 cm^3 of the phenylmagnesium bromide solution
(3.0 mmole).   The colour changes are similar to those observed for
the cobalt(II) complex except that the initial toluene suspension is
deep green and the product formed on exposure to air is pale green.

Suggestions for Further Reading

Principles of organometallic chemistry, G. E. Coates, M. L. H. Green, P. Powell and K. Wade. London: Methuen, 1968.

Organometallic compounds, G. E. Coates, M. L. H. Green and K. Wade, Third Edition, 2 volumes. London: Methuen, 1967 and 1968.

Advanced inorganic chemistry, F. A. Cotton and G. Wilkinson, Third Edition. New York: Wiley, 1972, especially chapters 23 and 24.

Organometallic chemistry, P. L. Pauson. London: Arnold, 1967.

Further references providing background reading to the discussion of the bonding are:

A valency primer, J. C. Speakman. London: Arnold, 1968.

Valency and molecular structure, E. Cartmell and G. W. A. Fowles, Third Edition. London: Butterworths, 1966.